The Racer's Guide To Fabricating Shop Equipment

By John Block

ISBN 0-936834-45-5

Editor: Steve Smith

Associate Editor: Georgiann Smith

Manufactured and printed in United States of America

Published by

 **STEVE SMITH AUTOSPORTS** PUBLICATIONS

P.O. Box 11631/Santa Ana, CA 92711/(714) 639-7681

ORDER HOTLINE (714) 639-7681

FAX LINE (714) 639-9741

- Shipping cost: Add $4 for first item, $7 for 2 items, $10 for 3 or more
- UPS delivery only. Provide street address
- Checks over $30 must clear first (30 days)
- Canadian personal checks not accepted
- Prices subject to change without notice
- In Canada add $5 to all orders
- CA residents add 7.75% sales tax
- US funds only. No COD's

STEVE SMITH AUTOSPORTS PUBLICATIONS

P. O. Box 11631-W
Santa Ana, CA 92711

Dedication

This book is dedicated to my wife, Jean. Her ideas, encouragement, and patience made this book a reality. The racing business, without extra projects, is quite difficult for one person. Thank you for helping me to achieve a dream.

John Block

Acknowledgments

A special thanks goes to Scotty Kerr for his garage and welding expertise on these projects. Thanks also goes to Steve Smith for his patience in seeing me through this book. I appreciate the help. Thanks guys.

John Block

INTRODUCTION

Racing requires a large amount of preparation and maintenance. Also, it takes more money than we would like to admit. Equipment for fabrication and repairs can reduce work time and racing costs, but the equipment can cost more than the racing.

This book provides a means for the average racer to build his own equipment. It covers everything from fabricating skills to the tools necessary to do the job. Each piece of equipment has assembly instructions, and operating instructions for the equipment once it is completed.

This equipment has been carefully designed. Prototypes of each piece have been built, tested, and evaluated. The designs and equipment shown in this book are the final result of an engineered program. Altering dimensions or substituting materials may result in structural failure and injury. Careless operation of any equipment can result in injury, so, please be careful.

Notice

The information contained in this book is true and accurate to the best of our knowledge. All recommendations and procedures in this book are made without any guarantees on the part of the author or Steve Smith Autosports. Please be aware that the use of the tools and equipment described in this book can be dangerous if used improperly, and injuries may occur. Because the use of the information in this book is beyond the control of the author and publisher, liability for such use is expressly disclaimed.

Table Of Contents

	Page
Dedication and Acknowledgements	3
Introduction and Notice	5
Fabricating Tips	9
Engine Stand	15
Hydraulic Press	21
Engine Hoist	29
Sheet Metal Brake	38
Flame Cutter	45
Chassis Stand	53

CHAPTER ONE

FABRICATING TIPS

Whether you are building race cars or equipment, certain fabricating skills are necessary. In this chapter we will cover the basic skills necessary to fabricate your new equipment. We will first cover buying materials. The necessary tools and equipment to do the work will follow. Next, general fabrication such as measuring, drilling, and cutting are covered. The final section in this chapter is devoted to welding basics and includes weld penetration and strength.

Buying materials

The first step in any project is to acquire the materials. You may already deal with a steel distributor for race car materials, but if not, check the phone book and call around for prices and availability. The materials used for the equipment in this book are common and should be readily available.

Metals come in different strengths. To be classified as structural steel the metal must have a yield strength of 36,000 pounds per square inch. Such material is graded as "A-36" steel. A-36 steel is a good material for our use because it can readily and safely be welded. Metals are also classified by the alloy, or mixture, of different types of materials in the metal. By choosing a certain alloy, specific properties can be achieved. "1010" mild steel alloy has been specified for our non-structural requirements. This is a mild steel which is relatively soft, easily formed, and welded. Use the materials that are specified! Choosing the wrong material can lead to structural failure and possible injury.

To reduce the cost of material, it is best to buy lengths. Most steel comes in a 12, 20, or 21 foot length which is called a "length." Metal suppliers will sell you less than a length, but often times charge a cutting fee. Design considerations for this equipment have been made to maximize the economy of buying full lengths. If you wish to build all of the equipment in the book, a material list follows which will minimize cost by buying in quantity:

A-36 Steel

20' - 2 1/2" x 2 1/2" x .120" square tubing
20' - 2" x 2" x 3/16" square tubing
20' - 2" x 2" x .120" square tubing
20' - 3 1/2" x 3 1/2" x 3/8" angle iron
20' - 4" x 5.4 lb channel
20' - 2" x 1" x 1/8" channel
21' - 1 1/4" x .145" pipe
24' - 2" x 1/8" CR flat stock
12' - 1" x 1/8" CR flat stock

1010 Steel

20' - 1 1/4" x 1 1/4" 16 ga. square

Other materials will be needed for the specific equipment so refer to the individual material lists.

Necessary tools and equipment

Certain hand tools and basic equipment are necessary for fabrication. Chances are that you already have most or all of the necessary tools and equipment. If not, the hand tools are inexpensive and should be in everybody's tool box. The larger equipment such as a drill press may be out of your budget, but chances are you know someone who will let you use theirs. The types of tools needed are grouped into fabricating requirements such as layout and drilling. Each tool is also listed with its intended use.

Layout tools

Tape measure and/or steel rule .... general measurements
Graduated caliper .... precision measurements
Scriber and/or prick punch .... drawing lines on steel
Pens and/or pencils .... drawing lines on patterns
Framing and/or combination square .... laying out lines and angles
Center punch and hammer .... laying out center dimples
"C" clamps and/or vise grips .... holding layout positions

Drilling tools

3/4" and 1/2" drill bits .... drilling specific holes
3/8" drill bit set .... drilling pilot holes
1/2" hand drill or drill press .... drilling
Round file and/or deburring knife .... deburring holes

Cutting tools

Cutting torch .... used in flame cutter
Band saw or chop saw .... cutting material
Grinder and/or belt sander .... deburring cut edges

Welding tools

AC arc welder or MIG welder .... welding connections
Slag hammer and wire brush .... cleaning and checking
Welding helmet and gloves .... protection

General Fabrication

Measuring

Measuring for layout is the first step in actual work. Accurate measurements are the basis of good work. The importance of careful measuring can mean the difference between a usable part and scrap. Measuring can be done with a tape or ruler. Because of the nature of transferring measurements from a ruler to the work, they are accurate to no more than 0.020". This is fine for most work, but sometimes more accuracy is required, in which case a "graduated caliper" is needed. Inaccuracy in measurements can happen with tape measures due to the little hook on the end of the tape. A simular problem occurs with the rounded ends on a ruler. To avoid such problems it is common to "burn an inch." To do this, start the measurement at the 1 inch mark rather than the hook or end. It is important to remember to add one inch to your measurement because you started at one rather than zero.

Layout Lines

Layout is the marking of lines for cutting or holes to be drilled. Layout is best done with a scriber or prick punch. Pencils and soap stones are okay, but they get dull and make a fat line which sometimes throws off the accuracy. Occasionally your marks will be hard to see. Layout dye can be used to accent scribe marks. My favorite is the spray dye. This dye is sprayed on the metal and then the lines are scribed through it. In a real pinch you can use spray paint, but it has a tendency to chip out in chunks, and not leave a fine line. If you use paint, spray it real thin so it won't chip. Squares are important for layout and should be used to assure that the lines are laid out "square" or at 90 degrees to the edge. A framing square or combination square will do the job.

Drilling Holes

Drilling holes is usually necessary in order build most any equipment. Always start by laying out the center of the hole. Next use a centerpunch and hammer to make a dimple at the center of the desired hole. This keeps the drill bit from walking off center when starting the drill. When using a drill press, before drilling the hole the bit should be lightly pressed on the work and then removed. The resulting impression can then be checked to see if the hole is being started in the desired location. With a drill press, never, never hold a piece of work in your hand. Clamp the work to the press table or lock it in a table vise. Also, remember to wear eye protection. When using a drill press it is possible to use a center drill to start a hole without the use of a layout dimple. The center drill can be used to center and countersink for a larger drill bit. Any time a large hole is going to be drilled, it should be pilot drilled. This is where a smaller hole is drilled first in order to guide the larger bit. The smaller bit should be close to the same size as the web of the larger bit. I always pilot drill any hole larger than 3/8-inch.

The secret to drilling a good hole is a sharp drill bit. A sharp bit will cut a clean, smooth and round hole. Once a bit has dulled it will squeal, or chatter, and not cut well. The most important factor in keeping a bit sharp is not to over heat it while drilling. Cutting oil should be put on the bit and dripped on the work while drilling. This helps carry away the heat and cool the bit.

Metal Cutting

Cutting metal can be done by a number of methods. Band saws, hack saws and cutting torches will all do the job. Cutting with a torch and then grinding the cut edge smooth will work, but it is slow and hard to cut accurate lengths. Cutting with a hack saw is not recommended on heavy material because it is slower yet. A band saw will do the job provided you have the proper blade. However, the best tool I've found is the chop saw. This is an abrasive wheel cut-off saw. This saw cuts quickly and leaves a clean, smooth edge. When using any saw it is important to make allowances for blade width. It's quite frustrating to cut a piece of material and then find out that it is a blade width short of the desired length. A good practice is to cut the longest desired lengths first. That way if you make a mistake the piece can be used somewhere else. Oftentimes a saw will leave a lip or burr on the edge of the cut. The cut edges should always be deburred to assure a proper fit between pieces. Safety with power saws is paramount. Be sure to keep your fingers clear of the blade and wear eye protection.

Arc Welding Tips

Safety is important with welding. Burns from welding are a common hazard. Flash burns are also a problem, this is where the light from the arc burns the retina of the eye. Be sure to use the proper helmet and lens. Gloves are also important. You can get a good sunburn from the ultraviolet light in the arc, so cover your skin.

Proper protective clothing is a must when arc welding. Use the proper helmet and lens, full heavy clothing to completely cover your skin, and gloves.

Most everyone has access to an arc welder or can afford to buy one. Arc welding is the process of fusing, or melting, metals together. The arc from an electrode causes the base metals to melt to create penetration. At the same time the base melts, the electrode also melts to form a bead. Penetration is the term used to indicate the depth from the original surface to the depth of the weld. Partial penetration is the failure of the bead to fuse with the metal part. Never weld over slag; it prevents penetration. No one can become a good welder by reading about welding. Welding skills come from practice. Also, to learn everything about welding from one chapter is impossible, but some important basics can be covered. In order to weld with the common AC stick welder, four basics must be first accomplished. These basics are striking an arc, rod position, arc length, and welding speed.

To begin welding, an arc must be established. Be sure the ground lead is making good electrical contact with the work before starting. With your helmet in place, scratch the electrode on the work as if striking a match. As you scratch the rod it should spark, indicating electrical contact. At that point, lift the rod about 1/8" and the arc will be established. If you stop moving the rod before the arc is established the rod will stick to the work. Most beginners try to start the arc by stabbing the rod, but they either stick the rod to the work or jerk it back too fast and break the arc. To begin, remember that you can't stab an arc, you must "strike" an arc.

Once the arc is established the welding rod must also be held in the proper position. It is best to weld with both hands. This gives you better control of the rod. To do this hold the "stinger" in your right hand (or left if left handed) and rest it in the palm of your other hand. It will also help to put your left elbow (or right if left handed) against your body. This provides extra support and you won't tire out so quickly. All this holding may sound awkward, but give it a try. If you are right handed you should weld from left to right whenever possible. This will help you see where you are going.

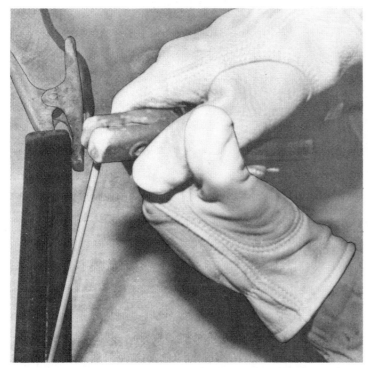

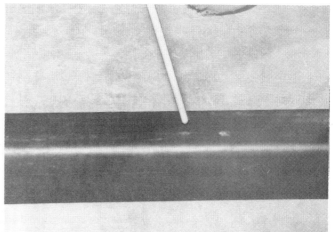

The rod should be positioned at a tilt of 15 to 20 degrees in the direction of weld travel to achieve a proper weld.

Once the arc is established the welding rod must be held in the proper position. It is much better to use two hands to control the weld.

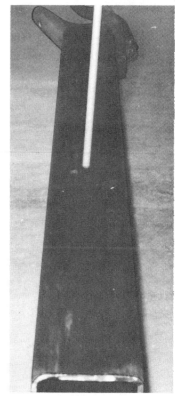

Now for the rod position. The position is important to help achieve the proper arc length. If you are welding from left to right, the top of the rod should be tilted to the right about 15 to 20 degrees. Which ever direction you are welding, the top of the rod should always be tilted in the direction of travel. At the same time the rod should be at a 90-degree angle to the work when viewed from the end.

The arc length is the distance from the end of the rod to the work. Once an arc has been established, maintaining the proper arc length is very important. Generally the arc should be short, approximately 1/16" to 1/8" long. An easy way to tell if the arc is the proper length is by the sound. A good arc has a crackling sound. A long arc has a open or hollow sound. As the welding takes place the rod will need to be fed into the work in order to maintain the proper arc length.

Welding speed is critical in establishing a good bead. The thickness of the material to be welded will be the main factor in determining the welding speed. Thinner metal must be welded faster than thicker metal. The thicker metal must be welded slower for good penetration. The correct speed can be determined by watching the molten puddle behind the arc. Do not watch the arc while welding. The appearance of the puddle and the ridge where the puddle becomes solid will be the indicator for the proper speed. The puddle

The rod should be held 90 degrees to the work piece as viewed from the end.

should flow and remain full of molten metal. The ridge where the puddle becomes solid should be around 3/8" behind the arc. Welding too fast will result in poor penetration. Also, the bead will not be a uniform width, often wiggling from edge to edge of the weld. Welding too slow will result in big blobs and often losing the arc. Generally, when welding at the proper speed it is not necessary to weave the arc.

The strength of a weld is mainly determined by the penetration. It is similar to the roots of a tree; without them the tree would fall over. The type of welding rod will also determine the strength of the weld. Don't worry about what type of rod should be used. With each piece of equipment and particular connection in this book, the type and size of rod will be specified, and the amperage setting for the weld.

CHAPTER TWO

ENGINE STAND

Materials List

8 1/2 ft. - 2" X 2" X .120" square tubing
4 in. - 3 1/2" I.D. X 3/16" pipe
4 1/2 in. - 3" I.D. X 3/16" pipe
10 in. - 3/4" X .120" round tubing
6" X 8" X 1/4" flat plate
86 in. - 1" X 1/8" CR flat stock
4 in. - 3/4" x 3/4" bar stock
1 ea. - 3/8" X 1" bolt
4 ea. - 1/2" X 3" bolt
4 ea. - 1/2" nut
8 ea. - 1/2" flat washers
3 ea. - 2" steel wheeled casters

Assembly Instructions

Step 1 Base/Stand

Lay out and cut the following lengths of 2" square tubing:

40 inches

38 inches

24 inches

Start with the longest length, then the next longest and so on. By doing this, if you make a mistake the piece can be used somewhere other than the scrap pile. Remember to make allowances for the saw blade width.

Step 2

Lay out a 3 1/2 degree angle on one end of the *24" length* of 2" square tubing. This angle can be quickly laid out by measuring 1/8" from the end on one edge and then drawing a line to the end of the opposite edge. Draw the 1/8" line across the adjacent side and lay out the same angle on the opposite face.

With a grinder, carefully remove the wedge formed by the 3 1/2 degree angle. Be sure that your new edge is smooth and even.

The other end of the *24" length* needs to be "fish-mouthed" for the 3 1/2" pipe. A fish-mouth is a notch where something round, like pipe, fits on another member. The fish-mouths should **not** be on the same face as the 3 1/2" degree angles. A fish-mouth can be laid out by using one end of the pipe as a template. To do this, place the pipe so the arc described by the pipe, runs from one edge to the other edge and meets the corners at the end. Double check to see that the fish-mouthed sides will not be on the same sides as the sides with the 3 1/2 degree angle.

With the grinder, remove the circular segment on both sides of the tubing to form the fish-mouth.

Step 3

On the *40" length* of 2" square tubing, lay out a line at 20" from the end.

On the *38" length* of 2" square tubing, lay out a centerline (approximately two inches long) at one end. A centerline marks the center axis of a member; in this case it would be one inch from either side and run long ways on the tubing.

Prepare to weld the *38" and 40" lengths* by placing the lengths together, aligning the centerline and 20" line, so that the two lengths form a "T." With a framing square check to be sure your placement is square.

Tack weld one corner of this connection. A tack weld is a small weld used to hold position before welding solid. Being careful not to break the tack weld, turn the work over. **Check for squareness** again, and then tack weld the opposite corner. Should the connection "pull" out of square from the tack weld, lightly tap the pieces back to square with a hammer.

Weld all four sides of this connection to make it solid. Use 1/8" 6013 welding rod and set the amperage at 105. The amperage may need to be altered if you're not getting good penetration. The work can be placed on its sides or back to facilitate flat horizontal welds, which are easier than vertical or overhead welding. Allow the weld to cool for a few minutes, then chip the slag away and inspect the weld. Any gaps or holes in the weld should be rewelded. With these pieces connected, this now forms the base.

Step 4

Lay out a centerline on the *24" length* in the same manner as the 38" length. This centerline should be on the end with the 3 1/2 degree angle. Draw the centerline on the long face, which is adjacent to the angled sides.

Prepare to weld the *24" length* to the base by placing it on the 40" length and aligning the centerlines of the 24" and 38" lengths. The 24" length should be standing upright and tilted

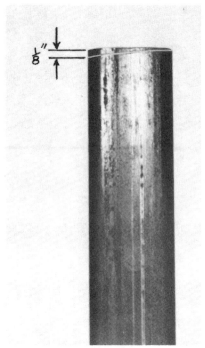

Step two, paragraph one

Step two, paragraph three

Step three, paragraph three

3 1/2 degrees away from the side with the 38" length. The fish-mouths should also be aligned with the 38" length. With a framing square, check for squareness between the 24" and 40" lengths.

Tack weld one corner of the connection, check the squareness and tack weld the opposite corner. Weld this connection solid and inspect as on the previous connection. This connection forms the upright.

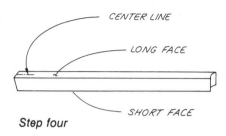

Step four

Step four, paragraph two

(Right) Step four, paragraph two, noting angle.

Step 5

Lay out and cut a *4" length* of 3 1/2" pipe if you don't already have it. On this piece, lay out a short line 2" from either end. With a center punch and hammer make a dimple in the line.

If you have a tap and die set, drill a 5/16" hole using the dimple as a guide for the drill bit. Once the hole is drilled use a 3/8" tap to thread the hole. If you thread the hole with a tap, go to step 6; if not, continue to the next paragraph.

If you do not have a tap and die set, drill a 3/8" hole instead. Place a 3/8" nut on a 3/8" bolt. Place the bolt through the hole from the outside of the pipe. Put a 3/8" nut, finger tight, on the bolt where it extends inside the pipe. Weld two sides of the nut which is on the outside of the pipe. Remove the nut inside the pipe and the bolt.

Step 6

Place the *4" length* of 3 1/2" pipe on the fish-mouthed upright. Orient it so that both ends extend 1" beyond the upright, and the bolt hole is upward. The centerline of the pipe should be aligned with the centerline of the 38" leg of the base.

Tack weld opposite corners of the fish-mouth to hold the position. With the pipe tacked in place, weld all four sides of this connection to make it solid. Use 1/8" 6013 welding rod and set the amperage at 105. The amperage may need to be altered if you're not getting good penetration.

Step 7

Lay out and cut the following lengths of 1" cold rolled flat stock:
8 @ 8 inches
2 @ 11 inches

Step six

Step 8

Lay out and cut opposing 45 degree angles on the ends of the two *11" lengths* of 1" flat stock. These angles can be laid out with a protractor-head square or by measurement. The measurement can be quickly laid out by measuring 1" from the end on one edge, and drawing a line to the end corner of the opposite edge. On the same side (edge) that you measured the 1", repeat the layout for the other end. These pieces will be used for braces.

On the 38" length leg of the base, on the same face as the centerline, lay out a line 4 1/4" from the upright. Lay out a line 4 1/4" from the base on the upright; this line should be on the face with the centerline.

Prepare to weld one brace in place by laying the stand on its side. Place the brace so that it forms a 45 degree angle running from the upright to the 38" leg of the base. Orient the opposing angled ends to be parallel with the edges of the base and upright. The brace should span the distance between the 4 1/4" marks on the upright and base. With the brace properly positioned, clamp it to the base and upright with "C" clamps.

Weld the brace solidly in place with 1/8" 6013 rod at 105 amps. Repeat the process with the other brace on the opposite side. Remove the slag from the weld and inspect for flaws.

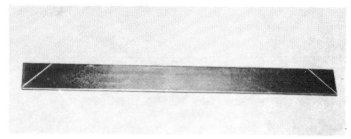

Step eight, paragraph one

Step eight, paragraph two

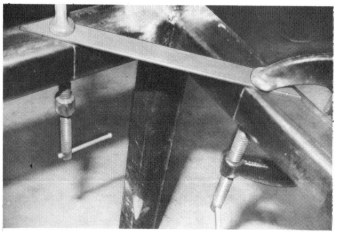

Step eight, paragraph three

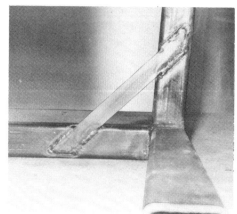

Step eight completed

Step 9

Turn the stand upside down and weld a caster at the end of each leg. Turn the stand right side up and place a 3/8" bolt in the top of the stand to lock the head in position. This now completes the stand

Step 10 Head Plate

Lay out and cut a 6" x 8" piece of 1/4" plate if you don't already have it. Use a grinder to round off any sharp corners. Lay out a line parallel with the edge so that the line is 1" from the edge. Repeat the lay out for the three other sides. With a center punch, make a dimple at each intersection of the four lines.

Drill four 1/2" holes using the dimples to guide the drill. With a file or deburring knife clean the edges of the holes.

Lay out the center of the *6" x 8" plate* by drawing a line 4" from one edge in the long direction, then drawing a line 3" from an adjacent edge. Lay out a mark 1 1/2" from the center on each line. This will help locate the 3" pipe in the middle of the plate.

Step ten

Step 11

Lay out and cut a *4 1/4" length* of 3" pipe if you don't already have it.

Place the *4 1/4" length* in the middle of the plate by using the layout marks. Tack weld opposite sides of the pipe to prevent moving, then weld solid using 1/8" 6013 welding rod at 120 amps. This now forms the head.

Place the 3" pipe of the head plate, inside the 3 1/2" pipe of the stand and lock in position with a 3/8" bolt.

Step 12 Fingers

Lay out and cut the 3/4" x 3/4" bar stock into the following lengths:

4 @ 1 inch

Place one of the pieces of 3/4" bar stock between two of the 8" lengths of 1" flat stock. Position the bar stock at the end of the flat straps so three sides of the bar stock are flush with the straps. With the pieces properly positioned, clamp them together with a "C" clamp.

Weld the bar stock and straps together with 1/8" 6013 rod at 90 amps. Repeat the process for the other three pieces of bar stock. Remove the slag from the weld and inspect for flaws.

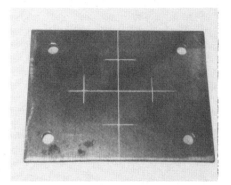

Step ten

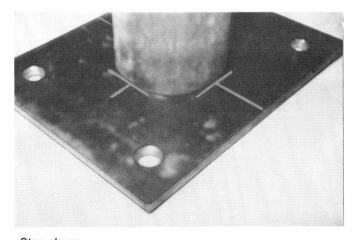

Step eleven

Step twelve

Step 13

Lay out and cut the 3/4" tubing into the following lengths:

4 @ 2 1/2 inches

Place a 2 1/2" length in the open end of a finger. Orient the tubing so that one end extends 1 3/8", and the other 1/8" beyond the edges of the 1" strap.

Weld the tubing in place with 1/8" 6013 rod at 90 amps. Remove the slag from the welds and inspect for flaws.

Repeat the process for the other three fingers.

Step 14

Place a 1/2″ bolt with a flat washer through the middle of each finger so that the bolt points in the opposite direction of the long side of the 3/4″ tube. Place another flat washer on the side of the bolt that sticks through the finger. Bolt each finger to the head with this bolt and washer configuration.

Step 15

You can now paint the engine stand for a completed project.

Things To Keep In Mind

Do not shorten the legs of the base or lengthen the upright. Altering dimensions can create instability and the stand will easily tip over. Neither furniture casters nor small rubber wheeled casters should be substituted for the casters in the materials list. Engines are heavy and these other types of casters will not carry the load.

When the head plate is properly positioned in the stand, approximately 1/4″ of the 3″ pipe is visible. Should the end of the 3″ pipe disappear into the 3 1/2″ pipe of the upright this is a warning that the head plate is working its way out.

Step thirteen

Step fourteen

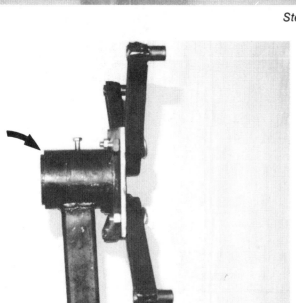

(Left) The head plate is properly positioned in the stand when approximately 1/4″ of the 3″ pipe is visible.

CHAPTER THREE

HYDRAULIC PRESS

Materials List

18 ft. - 2″ X 2″ X 3/16″ square tubing
5 ft. - 3 1/2″ X 3/8″ angle iron
5 ft. - 4″ X 5.4 lb. channel
5 ft. - 2″ X 1″ X 1/8″ channel
4 1/2 in. - 1 1/2″ X .145″ pipe
1 in. - 2″ round stock
2 in. - 1″ x 1/8″ flat stock
9 in. - 2″ x 1/2″ flat stock
2 ea. - 3/4″ X 3″ shouldered bolts and 3/4″ nuts
4 ea. - 3/4″ flat washers
2 ea. - 3/8″ X 3″ bolts and 3/8″ nuts

2 ea. - 10″ return spring
1 ea. - 10 ton hydraulic bottle jack

Assembly Instructions

Step 1 Main Frame

Lay out and cut the following lengths of 2" square tubing:

2 @ 60 inches
2 @ 30 inches
1 @ 24 inches
2 @ 2 1/32 inches
1 @ 2 inches

Start with the longest length, then the next longest and so on. By doing this, if you make a mistake, the piece can be used some place other than the scrap pile. Remember to make allowances for the width of the saw blade.

Step 2

On both of the *60" lengths* of 2" square tubing, lay out a line 25" from one end. From this line measure down and lay out four more lines 6 inches on center. It is important that the lines are exactly 6" on center, so double check your measurements. Now lay out a center line to cross each of the previous lines. With a center punch and hammer make a dimple at each of the five crosses. Without a drill press it will be necessary to repeat the process on the opposite face of the square tubing.

Drill a 3/4" hole at each dimple.

With a drill press, start by anchoring the work to the table so that when the drill penetrates the bottom side of the work it doesn't drill a hole in the table or vise. With a pilot drill, lightly press the bit on the work and quickly remove it. Check the resulting impression to see if the hole has started in the proper place. With the drill started in the correct location, drill through the work. Remove the the pilot drill and replace it with the larger drill bit. Now drill the large hole. Repeat the process for each of the dimples on both 60" lengths. A helpful hint: do not drill all of the pilot holes first and then come back to drill the big holes. This is because it is harder to recenter the drill bit once the work has been moved than to change the bit twice for each location. When finished, deburr the edges of the holes with a deburring knife or round file.

Without a drill press, drill a pilot hole at each of the dimples on both sides of the 60" lengths. Use the dimples to guide the drill bit. Change to the larger bit and drill all of the pilots to the larger size. Deburr the holes with a knife or round file. Hand drills are not as precise as a drill press. Therefore, when the work is assembled, touch-up filing may be needed on the holes.

Step 3

Measuring from the opposite end as the holes were measured from lay out, on both *60" lengths,* a line 7" from the end. The line should be on the same face of the tubing as the holes.

On the *24" length* of 2" square tubing, lay out a center line on both ends of the tube.

Prepare to weld the main frame together by systematically positioning and tack welding each piece in position. Begin by placing the *24" length* and one *60" length* of square tubing together; align the centerline of the 24" length with the line 7" from the end on the 60" length. The holes in the square tubing should be oriented up and not to the sides. With a framing square, check to be sure your placement is square.

Tack weld one corner of this connection. Being careful not to break the tack weld, turn the work over. Check for squareness again, and then tack weld the opposite corner. Should the connection "pull" out of square from the tack weld, lightly tap the pieces back to square with a hammer. Remove any slag or portion of the weld that would prevent the work from lying flat and turn the work over.

Place the remaining *60" length* together with the *24" length,* aligning the centerline of the 24" length with the line 7" from the end of the 60" length. With a framing square, check the squareness of the new placement. Measure the distance between the ends of the 60"

Step three

lengths. If the distance on both ends is not 24″, check to see that the *24″ length* is actually 24″ long. The distance between the ends should be the same. This is more important than the squareness of the connections.

Tack weld the *24″ and 60″ lengths* together using the same procedure as the previous connection. Remove any slag or portion of the weld that will prevent the frame from lying flat, and turn it back over.

Step 4

Lay out and cut the 3 1/2″ angle iron into two *28″ lengths.*

Lay out and cut opposing 45 degree angles on the ends of one face of both lengths.

On the opposite face of both *28″ lengths,* lay out a line 3″ from one end. Lay out another line 22″ from the first line. Measuring from the sharp edge of the angle (where both legs meet) lay out two lines 2 3/4″ from the edge to cross each of the previous two lines. Repeat this layout for the other 28″ length. With a center punch and hammer, make a dimple at the intersection of the layout lines.

Drill a 3/8″ hole at each dimple on both *28″lengths* of 3 1/2″ angle iron.

Step 5

On the *60″ lengths* lay out a line 3 1/2″ from the end opposite the 24″ length. The lines should be on the same face as the holes.

Place one of the *28″ lengths* of 3 1/2″ angle iron across the top of the 60″ lengths. Position it between the 3 1/2″ lines and the top. Orient the piece so the face with the 3/8″ holes is lying flat on the square tubing. The other face of the angle iron should be aligned with the 3 1/2″ lines. With "C" clamps or vise grips, clamp the piece into position. Place the clamps in the open ends of the square tubing so the frame can lie flat on the floor. Double check the measurement between the *60″ lengths.* If the measurements are not the same, tap the clamped ends to achieve equal distances.

Tack weld the angle iron to the square tubing and remove the clamps.

Turn the frame over and lay out a line 3 1/2″ from the ends, same as before. Position the remaining *28″ length* of angle iron in the same manner as the previous. Place a three inch long 3/8″ bolt through both holes in the *28″ lengths.* Check to be sure the piece is aligned with the layout lines. Now clamp into position.

Step five

Tack weld the angle iron and remove the clamps. Check the *60″ lengths* for flatness. To do this, place a straight edge across the face of both *60″ lengths.* Check to see that the face is flat with the straight edge. If one face touches the straight edge at a single point, break the necessary tack weld and flatten the *length.* Tack welds can easily be broken with a hammer and chisel.

Weld all connecting sides of the square tubing and the 3 1/2″ angle. Use 1/8″ 7018 welding rod and set the amperage at 120. Weld all four sides of the connections between the *24″ and 60″ lengths.* Use the same welding rod and amperage setting. The amperage may need to be altered if you're not getting good penetration. The work can be placed on its sides or back to facilitate flat horizontal welds, which are easier than vertical or overhead welding. Allow each weld to cool for a few minutes, then chip the slag away and inspect the weld. Any gaps or holes in the weld should be rewelded. With these pieces connected, this now forms the main frame.

Step 6

Lay out and cut opposing 45 degree angles on the ends of the two *30″ lengths* of 2″ square tubing. These angles may be laid out with a protractor-head square or by measurement. To lay out by measurement, on one end scale down 2″ from the end and draw a line to the end corner of the opposite edge. On the same side (edge) that you measured the 2″, repeat the lay out for the other end. When completed, these pieces will be used as the feet.

On both of the *30″ lengths* of square tubing that form the feet, lay out a line 15″ from either end. These lines should be on a face with the 45 degree angles.

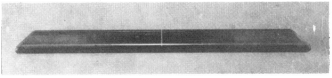

Step six

Step 7

On the bottom of the main frame (the opposite end from the 3 1/2" angle iron) lay out centerlines on the inside faces of the *60" lengths.* These centerlines should be on the face that connects the 24" length.

Stand the frame on its side. The centerline on the lower 60" length should now be up. Prepare to weld one of the feet to the main frame by placing these pieces together, aligning the centerline on the main frame with the line on the 30" length. This positioning should form a "T" shape. Orient the 45 degree angles to point toward the 3 1/2" angle. With a framing square, check to see that your placement is square, also that the main frame is square with the floor.

Step seven

Tack weld one corner of this connection, check the squareness and tack weld the opposite corner. Weld this connection solid using 1/8" 7018 rod with the amperage set at 120. Chip the slag and inspect for flaws.

Repeat the positioning and welding process for the other foot piece. When completed, stand the main frame upright on the feet.

Step 8

Lay out and cut a *9" length* of 2" x 1/2" flat bar, if you don't already have it. Lay out a centerline on the 2" face of the flat bar. Lay out a line 3 1/2" from one end to cross the centerline. Measuring in the same direction lay out another line 2" from the previous.

Lay out and cut two 2" lengths of 2 inch channel. Place the two pieces of channel on the flat bar. Orient the channels back to back and standing on one flange. Align the channels between the lay out lines and with the centerline between the two pieces. Properly positioned clamp the channel to the flat bar.

Weld the mid portion of each flange to the flat bar. Do not weld the ends of the channels to the flat bar. This will prevent the flat bar from sitting flat on the bottom of the 3 1/2" angle iron in the main frame. Weld the mid portion of the upper flanges together. Again, do not weld on the end of the flanges.

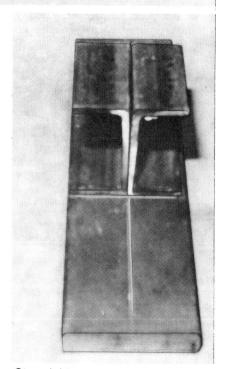

Step 9

Measure the ram diameter of the hydraulic jack you will use for this press. It will be necessary to have a piece of pipe with a slightly larger inside diameter than the ram's outside diameter. Probably a pipe with a 1 1/2" diameter will do, but check you jack first. Lay out and cut a *1/2" length* of pipe that will fit your jack.

Step eight

Lay out a line 4 1/2" from either end of the piece. Cross this line with a centerline. Place the *1/2" length* of pipe in the middle of the flat bar. Use the layout lines to center the pipe. Properly positioned, tack weld opposite sides of the pipe.

Weld the pipe to the flat bar with 1/8" 6011 welding rod at 95 amps.

Step 10

Lay out a line on the bottom faces of the 3 1/2" angle irons. This line should be 11" from one of the uprights. Lay out a second line 11" from the other upright.

Prepare to weld the flat bar to the main frame by positioning it between the lines previously drawn on the main frame. Orient the flat bar so the side with the 2" channel is between the two pieces of angle iron. Clamp into position and tack weld opposite sides of the flat bar.

Step ten

Weld the bar in position with 3/16" 6013 welding rod at 135 amps. Weld the upper flanges of the 2" channels to the angle iron with 1/8 6013 at 105 amps.

Step ten completed

Step 11 Work Bed

Lay out and cut the 4" channel into two 28" lengths.

Lay out and cut opposing 45 degree angles on the flanges at the ends of both lengths.

Lay out a line 1" from one end of both *28" lengths* of 4" channel. This line should be on the 4" face of the channel. Lay out a second line 26" from the first line for both channels. Now lay out a centerline to cross these lines. The centerlines should be 2" from both edges of the channel and running long ways. With a center punch and a hammer, make a dimple at the intersection of these lines.

Drill a 3/4" hole at each of the dimples on the *28" lengths.* Use the same drilling procedure as with the 60" lengths in the main frame. Remember to deburr the edges of the holes when finished.

Bolt both of the *28" lengths* of 4" channel to the main frame. Do this by placing a 3/4" bolt, with flat washer, through each hole from the flanged side of the channel. Place the piece in the main frame so the bolts pass through the holes in the *60" lengths.* Place the second *28" length* of 4" channel on the main frame so the bolts pass through the holes in the *28" length.* The 4" faces of the *28" lengths* should sandwich the *60" lengths* of square tubing in the main frame. Place a flat washer and 3/4" nut on the end of both bolts. With both 4" channels firmly bolted in place, they form the working surface. The five sets of holes in the main frame allow for five positions of the work bed.

Step 12 Jack Carriage

Lay out and cut the 2" channel into two 28" lengths.

Lay out and cut opposing 45 degree angles on the flanges at the ends of both lengths.

Lay out a line 2 1/32" from one end of both *28" lengths* of 2" channel. These lines should be on the 2" face of the channel. Lay out a second line 23 15/16" from the first line for both channels.

Step eleven

Prepare to weld the *2 1/32" lengths* of 2" square tubing to one of the 2" channels by placing each square tube adjacent to the layout lines. Orient the square tubes so an open end meets the 2" face of the channel. The square tubes should also be between the two layout lines so the measurement from the outside edges of the tubes equals 23 15/16". Once positioned, tack weld opposite corners of both square tubes.

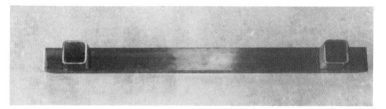

Prepare to weld the remaining *28" length* of 2" channel to the previous assembly by positioning the square tubes between the layout lines. The assembly should be placed on its side so as to be standing on the flange portion of the channels. This positioning is necessary to assure that the flanges will be flush with each other. Tack weld opposite corners of the square tubes to the unconnected 2" channel.

Weld the square tubes between the 2" channels with 1/8" 6011 rod at 105 amps. Weld only the portion of the square tube that meets the flanges of the channel. Do not weld on the 2" face of the channel. Remove any slag and inspect the weld. This now forms the carriage for the hydraulic jack.

Step 13

Lay out and cut a 4" length of 1 1/2" pipe if you don't already have it.

Step twelve

Insert the pipe into the *2" length* of square tubing. The pipe is approximately .003" larger than the opening in the square tubing, therefore the pipe will have to be driven into the square tubing with a hammer. To do this, grind or file a small radius on the edge of one end of the pipe. This will allow the pipe to start into the tubing. Place the square tubing open end up and start the pipe into the opening. Set a wooden block on the top of the pipe and drive the pipe into the tubing by striking the wood with the hammer. Do not drive the pipe by hitting it directly with the hammer; the end of the pipe must remain flat for another connection. The pipe should be driven into the square tubing until the ends of the pipe and tubing are flush.

Weld the four spots where the pipe and square tubing meet in the middle of the pipe. Use 1/8" 6013 rod at 105 amps. Do not weld on the flush ends of the pipe and tubing.

Prepare to weld the 2" round stock to the 1 1/2" pipe by standing the pipe on top of the round stock. Orient the pipe with the square tubing end up, and align the center of the pipe with the center of the round stock. Tack weld opposite sides of the pipe. Double check to be sure the pipe and round stock are centered with each other. When finished, this piece will be the ram. It is important that the ram be straight and concentric.

Weld the pipe and round stock together with 1/8" 6013 welding rod at 105 amps. This assembly will be used as the ram.

Step 14

Step thirteen

Prepare to weld the ram and the jack carriage together by positioning the jack carriage so both 2" channels are standing on the flange. Place the square end of the ram between the 2" channels. Center the ram between the two square tubes in the jack carriage. With a combination square, check the ram for squareness with the jack carriage.

Tack weld the square portion of the ram to the jack carriage. Turn the assembly over and check to be sure the other end of the ram is flush with the flanges of the jack carriage. Double check the squareness of the ram and jack carriage.

Weld the ram to the jack carriage along the 2" face of the channel in the carriage. Use 1/8" 6013 welding rod at 105 amps.

Step fourteen

Step 15

Lay out and cut the 1″ flat stock into two 1″ lengths. On both *1″ lengths* lay out a centerline, then lay out a line 3/8″ from one end. With a center punch and hammer, make a dimple at the intersection of the lines on both *1″ lengths.*

Drill a 1/4″ hole in both *1″ lengths,* using the dimple as a guide for the drill.

With a grinder, round off the corners of the end with the hole.

Lay out a centerline on both of the square tubes in the jack carriage. These lines should be on the side opposite the ram.

Prepare to weld one of the *1″ length* to the jack carriage by positioning the piece on the centerline just drawn on the square tubing. Orient the piece to be midway between the 2″ channels. It may be helpful to hold the piece in position with a pair of vise grips.

Weld the *1″ length* in position with 1/8″ 6013 welding rod at 95 amps. Repeat the process for the other 1″ length. This now forms the ears for the springs.

Step fifteen

Step 16

Hook one end of the return springs to the ears on the carriage. Place a 3/8″ bolt through the top of the main frame and the other end of the spring. Lock the bolt in position with a 3/8″ lock washer and nut.

Position the hydraulic jack on the carriage with the ram of the jack in the pipe at the top of the main frame.

Step sixteen

Step 17

You can now paint the press for a completed project.

Operating Instructions

*** WARNING ***

EXTREME CAUTION MUST BE USED, THIS PIECE OF EQUIPMENT CAN CAUSE SERIOUS INJURY OR POSSIBLY, DEATH.

This machine is rated at 10 tons. 20,000 pounds of force can shoot a piece of work across the garage, just like a bullet. Care should be used to insure the work is centered underneath the ram. The bolts through the 4" channels should also be tight. Safety glasses are a minimum requirement and a full face shield with safety glasses is better. OSHA requires hydraulic presses to be caged in the work place. Many serious accidents have happened while operating the hydraulic press. **Be careful**.

The hydraulic press can be very useful. The press is commonly used to remove or place bearings on an axle. This operation will require special "shoes" to hold the bearing while you are pressing the axle. Bearing shoes can be purchased or fabricated.

The press can also be used to bend or straighten materials. Punch and die sets can be used with a press to punch holes or special shapes. Whenever large amounts of force are needed to perform a job, your press and a little imagination can get it done.

CHAPTER FOUR
ENGINE HOIST

Materials List

20 ft. - 2 1/2" x 2 1/2" x .120 square tubing
22 in. - 2" x 2" square tubing
23 in. - 3 1/2" x 3 1/2" x 3/8" angle iron
13 1/2 ft. - 2" x 1/8" CR flat stock
6" X 27" X 1/4" flat plate
3 ea. - 3/4" x 3" shouldered bolt
4 ea. - 3/4" nuts and flat washers
4 ea. - 1/2" X 1" bolt

4 ea. - 1/2" nut
4 ea. - 3" steel wheeled casters
1 ea. - 6 ton hydraulic bottle jack

Assembly Instructions

Step 1

Lay out and cut the following lengths of 2 1/2″ square tubing:

2 @ 60 inches

1 @ 56 inches

1 @ 55 inches

Start with the longest length, then the next longest and so on. By doing this, if you make a mistake the piece can be used somewhere other than the scrap pile. Remember to make allowances for the saw blade width.

Lay out and cut the following lengths of 2″ square tubing:

1 @ 14 inches

1 @ 8 inches

Step 2 Base Assembly

Lay out and cut opposing 12 degree angles on the ends of the *14″ length* of 2″ square tubing. These angles may be laid out with a protractor-head square or by measurement. To lay out by measurement, on one edge scale down 7/16″ from the end and draw a line to the end corner of the opposite edge. On the same side (edge) that you measured the 7/16″, repeat the layout for the other end.

Lay out centerlines at both ends on both the long face and short face of the length.

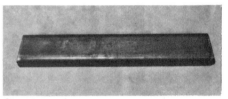

Step two

Step 3

Lay out and cut a 23″ length of 3 1/2″ angle if you don't already have it. Lay out and cut opposing 12 degree angles on one face of the *23″ length*. These angles may be laid out with a protractor-head square or by measurement. To lay out by measurement, on one face scale down 3/4″ from the end and draw a line to the end corner of the opposite edge. On the same side (edge) that you measured the 3/4″, repeat the layout for the other end.

Lay out two lines on the other face of the *23″ length* of angle iron. These lines should be 2 1/2″ from each end. Measuring from the sharp edge of the angle (where both legs meet) lay out two lines 7/16″ from the edge. Begin each line at the 2 1/2″ layout line and extend it to the end.

Cut out the 2 1/2″ wide rectangle on both ends of the 3 1/2″ angle iron. With the rectangles removed, grind the cut edges smooth. A lip may have been left on the inside of the face with the 12 degree ends. This face should be flat; therefore, any lip must be ground flat with the rest of the face.

Step three

Step 4

Lay out a line in the middle of the 3 1/2″ angle iron on the face with the 12 degree angles. Lay out two lines 2″ on each side of the previous line. Measuring from the sharp edge of the angle iron, lay out a line 1 1/2″ from the edge to cross the two previous layout lines. With a center punch and hammer, make a dimple at the intersection of the outer two layout lines.

Drill a 1/2″ hole at each dimple.

With a drill press, start by anchoring the work to the table so that when the drill penetrates the bottom side of the work it doesn't drill a hole in the table or vise. With a pilot drill, lightly press the bit on the work and quickly remove it. Check the resulting impression to see if the hole has started in the proper place. With the drill started in the correct position, drill through the work. Remove the pilot drill and replace it with the larger drill bit. Now drill the large hole. Repeat the process for the other dimple on the angle iron. A helpful hint: do not drill one pilot hole after another and then come back to drill the big holes. This is because it is harder to recenter the drill bit once the work has been moved than to change the bit twice for each location. When finished, deburr the edges of the holes with a deburring knife or round file.

Without a drill press, drill a pilot hole at each dimple on the angle iron. Use the dimples to guide the drill bit. Change to the larger bit and drill the pilots to the larger size. Deburr the holes with a knife or round file.

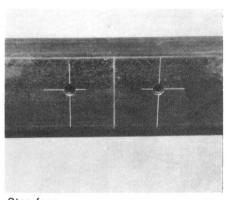

Step four

Step 5

Lay out a line 1" from the end for both of the *60" lengths* of 2 1/2" square tubing. Lay out a centerline to cross the 1" layout line on both pieces. With a center punch and hammer, make a dimple at the intersection of the layout lines.

Drill a 1/2" hole at each dimple, but through one wall only. Use the dimple to center the drill bit. Remember to deburr the holes when finished.

Step 6

Lay out three lines on both of the *60" lengths* of 2 1/2" square tubing. These lines should be laid out at 3", 5 1/16", and 15 5/8" from the end with the hole. Lay out a centerline, on both lengths, to cross the 3" and 5 1/16" lay out lines.

Prepare to weld the *14" length* of 2" square tubing to one of the *60" lengths* by placing the end of the 14" length between the 3" and 5 1/16" layout lines of the 60 length. Align the centerline of the 60" length with the centerlines on end of the 14" length. The *14" length* should be oriented so the piece tilts toward the end of the 60" length with the 1/2" hole.

Tack weld one corner of this connection. A tack weld is a small weld used to hold position before welding solid. Tack weld the opposite corner of the connection. Should the connection "pull" from the tack weld, lightly tap the pieces back together with a hammer.

Place the remaining *60" length* at the opposite end of the *14" length* as with the previous piece and tack weld in position. When completed, this assembly will be the base.

Prepare to weld the 3 1/2" angle iron to the base assembly by placing it between the 60" lengths. Align the outside of the face with the 2 1/2" rectangles with the 15 5/8" layout lines on the 60" lengths. Properly positioned, the 12 degree angles of the angle iron should be flush with the outside edges of the 60" lengths, also the face with the 2 1/2" rectangles should contact the inside of the 60" lengths. Tack weld both ends of the 3 1/2" angle iron to the base assembly.

Weld all four connections of this assembly to make it solid. Use 1/8" 6013 welding rod and set the amperage at 135. The amperage may need to be altered if you're not getting good penetration. The work can be placed on its sides or upside down to facilitate flat horizontal welds, which are easier than vertical or overhead welding. Allow the weld to cool for a few minutes, then chip the slag away and inspect the weld. Any gaps or holes in the weld should be rewelded.

Step 7

Weld a caster to the bottom side at the end of each *60" length*. With these pieces connected, this now forms the base.

Step 8 Upright Assembly

Lay out and cut a 14 degree angle on one end of the *55" length* of 2 1/2" square tubing. This angle may be laid out with a protractor-head square or by measurement. To lay out by measurement, on one edge scale down 5/8" from the end and draw a line to the end corner of the opposite edge.

Lay out three lines on the *55" length* of 2 1/2" square tubing. These lines should be laid out at 2 1/2", 12 1/2", and 24 7/8" from the square end and on both faces with the 14 degree angle on the opposite end.

Step 9

Lay out and cut two 6" x 2 1/2" pieces of 1/4" flat plate. With a grinder, remove any slag and straighten each cut edge.

Lay out a line 1" from either end of both plates. Lay out a centerline to intersect with the previous layout line on each plate. With a center punch and hammer, make a dimple at the intersection of the layout lines on both plates.

Drill a 3/4" hole at each dimple. Remember to deburr the holes when finished.

Step five

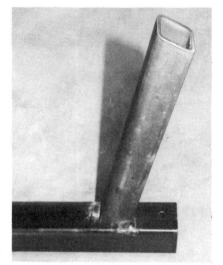

Step six

Step six

Step six

Step seven

Step nine

Step nine

Step nine

With a grinder, round the corners of the plates on the end with the hole.

Prepare to weld one of the flat plates to the *55" length* of tubing by placing the plate at the square end of the tubing. Align the solid end of the plate with the 2 1/2" layout line on the tubing. The end of the plate with the hole should extend beyond the end of the tubing. With the plate properly located, clamp it to the tubing with a "C" clamp. Tack weld the plate to the tubing at two places and remove the clamp.

Position the remaining plate on the square tubing in the same manner. Place a 3/4" bolt through the holes of both plates to assure proper alignment. With this plate properly located, clamp it to the tubing with a "C" clamp.

Weld the plates to the tubing with 1/8" welding rod with the amperage set at 135. Remove the "C" clamp and 3/4" bolt

Step 10

Lay out and cut two pieces of 1/4" plate as dimensioned in drawing. A helpful hint is to use a framing square to lay out the 3" and 12" side simultaneously. With a grinder, remove any slag and straighten each cut edge.

On each plate lay out a line 1" from the 3" long edge. Lay out a line 1" from the 12" long edge to cross the previous line. With a center punch and hammer, make a dimple at the intersection of the layout lines.

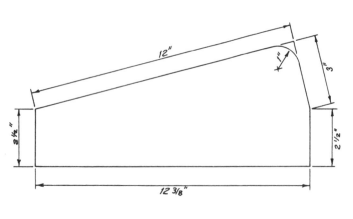

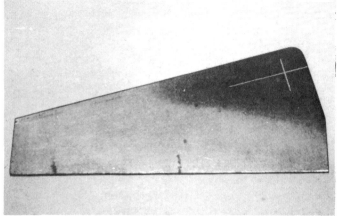

Step ten

Drill a 3/4" hole at the dimple on each of the two plates.

Prepare to weld one of the flat plates to the *55" length* of tubing by placing the plate on the the tubing between the 12 1/2" and 24 7/8" layout lines. The end of the plate with the hole should be oriented nearest the 12 1/2" layout line. Properly located, clamp the plate to the tubing with a "C" clamp. Tack weld the plate to the tubing at two places and remove the clamp.

Position the remaining plate on the square tubing in the same manner as the previous plate. Place a 3/4" bolt through the holes of both plates to assure proper alignment. With the plate properly located, clamp it to the tubing with a "C" clamp.

Weld the two plates to the square tubing with 1/8" 6013 welding rod at 135 amps. Remove the "C" clamp and 3/4" bolt.

Step 11

Lay out and cut a 3" x 5 1/2" piece of 1/4" flat plate if you don't already have it. Lay out two lines on the plate 3/4" from each end. Lay out a centerline to cross each of the previous lines. With a center punch and hammer, make a dimple at the intersection of the center-line and the two layout lines.

Drill a 1/2" hole at the two dimples on the 1/4" flat plate.

Step 12

Lay out two lines 1 1/2" from each end of the 1/4" plate.

Prepare to weld the 1/4" plate to the *55" length* of 2 1/2" square tubing by placing the angled end of the tubing on the plate between the 1 1/2" layout lines. Orient the tubing so the faces with the angles are adjacent to the 1/2" holes. Properly positioned, tack weld the plate to the tubing.

Bolt this assembly to the base. Do this by placing two 1/2" bolts through the 1/4" plate and 3 1/2" angle iron. Orient the *55" length* to tilt toward the narrow end of the base. With a framing square, check the the squareness between the upright and the angle iron.

Weld the plate to the tubing with 1/8" welding rod at 1200 amps.

Step 13

Lay out and cut the following lengths of 2" x 1/8" flat stock:

2 @ 55 inches

1 @ 52 inches

Remember to cut the longer pieces first.

Step 14

Lay out a line 1" from the end of both *55" lengths* of flat stock. Lay out a centerline on each piece to cross the previous line. With a center punch and hammer, make a dimple at the intersection of the layout lines on both pieces.

Drill a 1/2" hole at the dimple on both pieces. Remember to deburr the holes when finished.

Step 15

Lay out lines 2" from each end of both *55" lengths* of 2" flat stock.

Place one end of either piece in a vise. Orient the piece so the 2" layout line is approximately 1/16" outside the jaws and tighten the vise. The flat stock must now be bent 5 1/2 degrees at the 2" layout line. To accomplish this, clamp a piece of angle iron to the flat stock. The angle iron should be about two feet long with a cross section no smaller than 1" x 1" x 1/8". The angle iron stiffens the flat stock and allows the bend to occur at the layout line. Orient the length of the angle iron against the flat stock with one end approximately 1/16" from the layout line at the vise jaw. Press on the angle iron to bend the flat stock. A

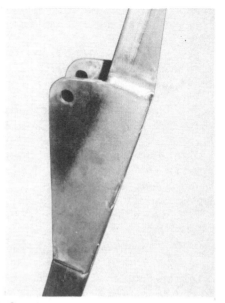

Step ten

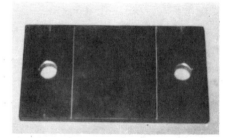

Step twelve

Step twelve completed

protractor will be needed to identify the proper bend. Be careful not to bend the material back and forth to achieve the desired angle. This will weaken the material and be a point of potential failure.

Bend the flat stock 5 1/2 degrees at the other layout line, but in the opposite direction as the previous bend. It may be easier to place the short end in the vise to bend the material, but be careful to bend the proper direction.

Repeat the bending process for the remaining *55" length* of flat stock.

Bolt both of the *55" lengths* of 2" flat stock to the base with 1/2" bolts. The free ends of the 55" lengths are to connect to the top of the upright. It will be necessary to slightly twist each 55" length to properly connect to the upright. To twist these pieces, grasp the free end with a large crescent wrench. Twist each piece so the 2" portion at the end will lie flat against the 1/4" plate at the top of the upright. When properly fitted, clamp the free ends of the *55" lengths* to the top of the upright. Tack weld both pieces of flat stock to the 1/4" plate on top of the upright.

Weld the flat stock to the upright with 1/8" rod at 120 amps.

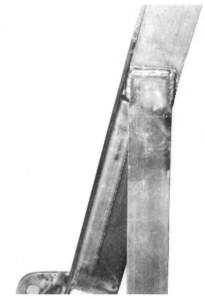

Step fifteen, paragraph six

Step 16 Beam Assembly

Lay out a line 2" from one end of the *56" length* of 2 1/2" square tubing. Lay out a line 1" from the opposite end. Lay out a centerline at each end of the tubing to cross the two previous layout lines. With a center punch and hammer, make a dimple at the intersection of the layout lines. Without a drill press it will be necessary to repeat the layout for the opposite face of the square tubing.

Drill a 3/4" hole at the dimple on the 2" layout line. Drill a 1/2" hole at the dimple on the 1" layout line.

Lay out a line 12" from the end with the 3/4" hole. This line should be drawn on either face, of the beam, without the holes. Lay out a centerline to cross the previous layout line. Also lay out a centerline approximately 8" from the end with the 1/2" hole.

Step 17

Lay out and cut an 8" length of 2" square tubing if you don't already have one. Lay out a centerline on each of the four faces at one end.

Prepare to weld the *8" length* of 2" square tubing to the *56" length* of 2 1/2" square tubing by orienting the 12" layout line on the 56" length upward. Stand the 2" tubing on the 2 1/2" tubing at the layout lines. Align the centerlines of the 2" tubing with the centerline and 12" layout line of the 2 1/2" tubing. When properly positioned, tack weld the two pieces together.

Weld the 2" tubing to the 2 1/2" tubing with 1/8" 7014 welding rod at 90 amps.

Step seventeen

Step 18

Lay out four lines on the *52" length* of 2" flat stock. Measuring from one end, lay out the lines at 2", 14", 16", and 50". Lay out a centerline at the end near the 50" layout line.

Place the end with the 2" layout line in a vise. Orient the piece so the 2" layout line is approximately 1/16" outside the jaws and tighten the vise. The flat stock must now be bent 41 1/2 degrees at the 2" layout line. To accomplish this, clamp a piece of angle iron to the flat stock. Orient the length of the angle iron against the flat stock with the end approximately 1/16" from the layout line at the vise jaw. Press on the angle iron to bend the flat stock. A protractor will be needed to identify the proper bend. Be careful not to bend the material back and forth to achieve the desired angle. This will weaken the material and be a point of potential failure.

Bend the flat stock 41 1/2 degrees at the 14" layout line, but in the opposite direction as the previous bend.

Bend the flat stock 13 1/2 degrees at the 16" layout line. The bend should be in the same direction as the last bend.

Bend the flat stock 13 1/2 degrees at the 50" layout line, but in the opposite direction as the previous bend. It may be easier to place the short end in the vise to bend the material, but be careful to bend the proper direction. When completed this piece will be used as a truss for the hoist beam.

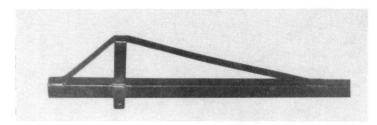

Step eighteen

Prepare to weld the flat stock to the beam assembly by placing the piece on top of the 2″ square tubing and 2 1/2″ square tubing. Some minor alterations may need to be made to the flat stock before it will fit properly on the beam. The 2″ portion at each end of the flat stock should lie flat on the 2 1/2″ tubing. The portion crossing the open end of the 2″ tubing should also be flat. Align the edges of the flat stock with the 2″ tubing, being sure the centerline on the flat stock is aligned with the centerline on the 2 1/2″ tubing. Attach a ″C″ clamp to the beam, clamping the flat stock to the top of the 2″ tubing. Tack weld the flat stock to the 2″ tubing on two sides and remove the clamp. Place a ″C″ clamp at the end of the flat stock with the centerline and clamp the flat stock to the 2 1/2″ tubing. Tack weld the flat stock to the beam at two places and remove the clamp. Finally, clamp the remaining end of the flat stock and tack weld it to the beam.

Weld the flat stock to the beam assembly with 1/8″ 7014 rod at 90 amps.

Step 19

Lay out and cut two 2″ x 4 1/2″ pieces of 1/4″ flat plate. With a grinder, remove any slag and straighten each cut edge.

Lay out a line 1″ from either end of the 1/4″ flat plate. Lay out a centerline to intersect with the previous layout line. Repeat the layout procedure for the remaining piece of 1/4″ flat plate. With a center punch and hammer, make a dimple at the intersection of the layout lines on both plates.

Drill a 3/4″ hole at the dimple on each plate.

With a grinder, round the corners of the plates at the ends with the hole.

Step 20

Lay out two lines on the side of the beam assembly. These lines should be at 11″ and 13″ from the end with the 3/4″ hole. Repeat the lay out for the opposite face of the beam.

Prepare to weld one of the flat plates to the beam by placing the plate between the layout lines on the side of the beam. Orient the plate so the end with the hole extends past the edge of the beam and the opposite end is flush with the surface where the 2″ tubing contacts the beam. When properly positioned, clamp the plate to the beam and tack weld in position and remove clamp. Repeat the placement for the remaining plate on the opposite side of the beam. With the second plate properly positioned, place a 3/4″ bolt through the holes of both plates to assure alignment. Clamp the plate to the beam.

Weld the plates to the beam assembly with 3/16″ 6013 rod at 120 amps. Remove the clamp and 3/4″ bolt.

Install the beam assembly to the upright with a 3 1/2 inch long 3/4″ shouldered bolt. Do this by placing the end of the beam with the 3/4″ hole between the two 1/4″ plates on top of the upright. Place the 3/4″ bolt through the plates and beam assembly. Place a 3/4″ nut on the end of the bolt, but not tight against the plate of the upright. Place a second nut on the bolt and tighten it against the first nut.

Step twenty

Step 21

Lay out and cut a 2″ length of 1″ pipe. The pipe should be just large enough to fit over the end of the jack ram. If your jack has a larger ram, a larger pipe will be needed. Lay out a line 5/8″ from one end. With a center punch and hammer, make a dimple on the line.

Drill a 3/4″ hole through both walls of the pipe using the dimple to center the drill with the pipe.

(Left and right) Steps sixteen through twenty-two

Install the pipe on the beam with a 3 1/2 inch long 3/4" shouldered bolt. Do this by placing the end of the pipe with the 3/4" hole between the two 1/4" plates on the beam. Place the 3/4" bolt through the plates and pipe. Use 3/4" flat washers to center the pipe between the plates. Place a 3/4" nut on the end of the bolt but not tight against the plate on the beam. Place a second nut on the bolt and tighten it against the first nut.

Step 22 Jack Perch

Lay out and cut a 4 1/4" x 4 1/4" piece of 1/4" flat plate. The jack will sit on this piece. If your jack has a larger base, the plate will need to be larger. With a grinder, remove any slag and straighten each cut edge.

Lay out a line 9/16" from the edge of the 4 1/4" x 4 1/4" plate. Lay out another line 3 1/8" from the first line.

Step 23

Lay out and cut two 2" x 4" pieces of 1/4" flat plate. With a grinder, remove any slag and straighten each cut edge.

Lay out a line 2" from either end of both plates. Lay out a line 1 3/16" from the long edge to cross the previous layout lines. With a center punch and hammer, make a dimple at the intersection of the layout lines on both plates.

Drill a 3/4" hole at the dimple on both plates. Remember to deburr the holes when finished. These pieces will be used as flanges for the jack perch.

Prepare to weld the flanges to the 4 1/4" x 4 1/4" plate by standing the flanges on their 4" sides adjacent to each layout line. The flanges should be on the outside of the layout lines. Place a 3/4" bolt through the holes in the flanges to assure alignment. When properly positioned, tack weld each flange to the plate and remove the 3/4" bolt.

Weld the flanges to the plate with 1/8" 6013 rod at 120 amps. Remove any slag and inspect the welds.

Install the jack perch on the upright with a 3/4" bolt. Do this by placing the bolt through one flange of the perch, the two 1/4" plates in the middle of the upright, and the second flange of the perch. Place a 3/4" nut on the bolt and thread it to the flange. With a wrench, lightly snug the 3/4" nut.

Step 24

Place the jack on the perch with the ram extending into the pipe on the bottom of the beam assembly.

A coat of paint will provide a nice finish for your new hoist.

Step twenty-two

Step twenty-three

Operating Tips

This piece of equipment is for lifting motors or other heavy objects. Do not use it to transport an object.

Do not alter the dimensions of this piece of equipment. Altering dimensions or substituting materials may lead to structural failure.

NOTE: Hoist capacity is 1,000 pounds maximum. DO NOT use it to cart engines all over your shop.

The finished product

CHAPTER FIVE

SHEET METAL BRAKE

Materials List

10 ft. - 4″ X 5.4 lb. channel
137 in. - 2″ X 1″ X 1/8″ channel
10 ft. - 3 1/2″ X 3/8″ angle iron
58 in. - 1 1/4″ x .145″ pipe
20 in. - 1″ x .120 tubing
4 in. - 3/4″ round stock
8 in. - 1/4″ round stock
14 1/2″ X 8 1/4″ X 1/4″ flat plate
2 ea. - 3/8″ x 3″ shouldered bolts
2 ea. - 3/8″ x 3″ bolts
8 ea. - 3/8″ nuts
2 ea. - valve springs
        (any common spring not taller than 2 1/2″)
        (used springs are fine)

Assembly Instructions

Step 1 Base/Stand

Lay out and cut the following lengths of 4" channel:

1 @ 60 1/8 inches
2 @ 30 inches

Start with the longest length, then the next longest and so on. By doing this, if you make a mistake, the piece can be used some place other than the scrap pile. Remember to make allowances for the width of the saw blade.

Step 2

Lay out and cut opposing 45 degree angles on the ends of the two *30" lengths* of 4" channel. These angles may be laid out with a protractor-head square or by measurement. To lay out by measurement, orient the channel with the flat side up, on one end scale down 4" from the end and draw a line to the end corner of the opposite edge. On the same side (edge) that you measured the 4", repeat the lay out for the other end. When completed these pieces will be used as the feet.

On the short flange of each foot piece, lay out a line 10" from the end of the flange. Be careful to measure from the end of the flange face and not the underside of the flange. Measuring in the same direction as before, lay out a second line 2" from the previous line.

Step 3

Lay out and cut the following lengths of 2" channel:

1 @ 60 1/8 inches
2 @ 36 inches
2 @ 2 inches

Step 4

Prepare to weld a *36" length* of 2" channel to one of the feet by placing these pieces together to form a "T" shape. Orient both channels so the flanges are upward. The end of the 2" channel should be between the two layout lines on the foot. With a framing square, check to be sure your placement is square.

Tack weld opposite ends of this connection. A tack weld is a small weld used to hold position before welding solid. Check for squareness again before welding solid. Should the connection "pull" out of square from the tack weld, lightly tap the pieces back to square with a hammer.

Weld all sides of this connection to make it solid. Use 1/8" 6013 welding rod and set the amperage at 120. The amperage may need to be altered if you're not getting good penetration. The work can be placed on its sides or back to facilitate flat horizontal welds, which are easier than vertical or overhead welding. Allow the weld to cool for a few minutes, then chip the slag away and inspect the weld. Any gaps or holes in the weld should be rewelded. This now forms one side of the base stand.

Repeat the positioning and welding process for the other *36" length* of 2" channel and foot piece. When completed, these pieces form the sides of the base stand.

Step 5

Lay out and cut a 58" length of 1 1/4" pipe if you don't already have it.

Prepare to weld the *58" length* to both sides of the base by standing the sides upright and orienting the flanges of the channels toward each other. Position these sides approximately 58" apart. Place the pipe to rest on the top flange of the feet and fit inside the 2" channel. The ends of the pipe should be against the inside face of the 2" channels on both sides. With a framing square, check the squareness between the pipe and 2" channel uprights.

Step four

Step five

Tack weld the pipe in two spots on both feet. Check the measurement at the top of the uprights. If the dimension is not 58″, lightly tap the necessary upright to achieve the proper distance.

Weld the pipe solidly into position with 1/8″ 6013 welding rod at 120 amps. Remove any slag and inspect the welds. This now forms the base for the sheet metal brake.

Step 6 Brake Assembly

Lay out and cut a cardboard template as dimensioned in drawing (A). Place the template on the 1/4″ flat plate and trace out the template with soap stone. Move the template and trace it again.

With a cutting torch, cut the two tracings out of the 1/4″ plate. These pieces are going to be used as end plates.

With a grinder, remove any slag left on the end plates from the cutting torch. Lay one plate on top of the other and align the edges of the top and bottom plate as best as possible. Clamp the plates together with two vise grips. One edge at a time, grind both plates until the edges of both plates are smooth and uniform. The vise grips will need to be relocated in order to finish certain sides. When moving the vise grips, unclamp, move, and reclamp the grips one at a time to prevent slippage between the plates. Use a framing square to check the squareness of the corners. With each side straight and smooth, radius the corners of the plates with a grinder.

Step 7

Lay out lines on one end plate as shown in drawing (B). Repeat the layout for the other end plate, but on the opposite face as the first plate.

The drawing shows two line that intersect in the center of a dotted circle. With a center punch and hammer, make a dimple at the intersection of these lines on each plate.

Drill a 3/4″ hole at each dimple.

With a drill press, start by anchoring the work to the table so that when the drill penetrates the bottom side of the work it doesn't drill a hole in the table or vise. With a pilot drill, lightly press the bit on the work and quickly remove it. Check the resulting impression to see if the hole has started in the proper place. With the drill started in the correct position, drill through the work. Remove the the pilot drill and replace it with the larger drill bit. Now drill the large hole. Repeat the process for the dimple on the other end plates. A helpful hint is not to drill one pilot hole after another and then come back to drill the big holes. This is because it is harder to recenter the drill bit once the work has been moved than to change the bit twice for each location. When finished, deburr the edges of the holes with a deburring knife or round file.

Without a drill press, drill a pilot hole at the dimple on both end plates. Use the dimples to guide the drill bit. Change to the larger bit and drill the pilots to the larger size. Deburr the holes with a knife or round file.

Step five

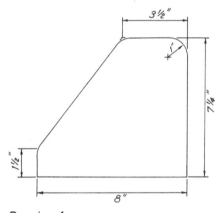

Drawing A

Step seven

(Right) This is the end plate you are fabricating.

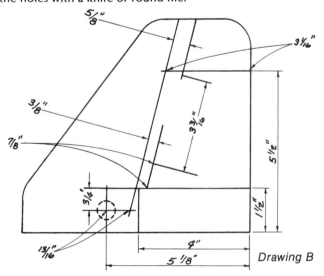

Drawing B

Step 8

Lay out a line 1/2" from the end of both *2" lengths* of 2" channel. These lines should be on the 2" face of the channel. On one piece lay out a second line 3/8" from the left side. On the remaining piece lay out a second line 3/8" from the right side. With a center punch and hammer, make a dimple at the intersection of these lines.

Drill a 3/8" hole in each *2" length* of 2" channel.

Prepare to weld the *2" length* of channel to the end plate by positioning the length adjacent to the angular layout lines as seen in the photo. Orient the channel so the end with the hole is upward. Be sure the hole is not near the side that is adjacent to the layout line. Tack weld the piece into position; weld only on the inside of the 2" face. When completed, this will serve as the spring perch.

Weld the the spring perch and end plate together with 1/8" 6011 welding rod at 90 amps. Do not weld on the face of the flange adjacent to the layout line. Repeat the positioning and welding process for the remaining end plate and spring perch.

Step 9

Prepare to weld the *60 1/8" length* of 4" channel to one end plate by standing the 4" channel on the end plate with the channel resting on its end. Align the channel with the layout lines on the end plate as seen in the photo. With a framing square, check the squareness of the placement of the channel on the end plate. Tack weld the channel into position; weld only the flange side of the 4" face.

Prepare to weld the *60 1/8" length* of 2" channel to one end plate by standing the 2" channel on end and aligning it with the layout lines on the end plate as seen in the photo. With a framing square, check the squareness of your placement. Tack weld two sides of the channel to hold its position.

Prepare to weld the remaining end plate to this assembly by carefully turning the assembly over and standing the channels on the unconnected end plate. Align the 4" channel with the layout lines as before, and tack weld into position. Again, weld only the flange side of the 4" face. Now align the 2" channel with the layout lines and tack weld it into position.

Weld the 4" channel to the end plates with 1/8" 6013 rod at 135 amps. Do not weld on the 4" face or the flange face adjacent to the 3/4" hole.

Weld the 2" channel to the end plates with 1/8" 6013 rod at 120 amps. Remove any slag and inspect the welds for both channels.

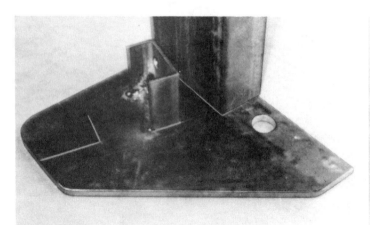

Step nine

Step 10

Layout and cut two 2" lengths of 3/4" round stock.
Layout and cut the 3 1/2" angle iron into two 60" lengths.

Step 11

Lay the brake assembly flat, resting it on the 7 1/4" sides of the end plates. This assembly is bulky and heavy, therefore, it is a good idea to have help moving it around. Place one of the *60" lengths* of angle iron into the assembly with one leg (side) flat on the ground and the other pointing upward and against the 4" channel. Place a 2" length of 3/4" round stock in the hole of each end plate. Orient the end of the round stock to be flush with the outside face of the end plate. Orient the angle iron to contact the round stock with both legs. To do this place shims under the angle iron so the round stock is will lie level on the angle iron.

Tack weld the round stock to the angle iron. Check to see that the 3 1/2" angle pivots freely in the assembly without binding. Weld both *2" lengths* of round stock solid to the angle with 1/8" 6013 welding rod with the welder set at 120 amps.

Step eleven

Step 12

Place the brake assembly on top of the stand. Orient the brake assembly so the 4" channel rests on the stand. The uprights of the stand should be approximately 1" from the end plates. With the assembly properly positioned, tack weld both uprights to the underside of the brake assembly.

Weld the brake assembly to the stand with 1/8" 6011 welding rod at 120 amps. Remove any slag and inspect the welds.

Step 13 Upper Blade

Lay out a line 1 7/16" from the end of the remaining *60" length* of 3 1/2" angle iron. Repeat the layout for the other end of the 60" length. Measuring from the sharp edge of the angle (where both legs meet) lay out two more lines 2" from the edge. These lines should cross each of the previous lines. With a center punch and hammer, make a dimple at the intersection of the layout lines.

Step twelve

Drill a 3/8" hole at each of the dimples on the *60" length*. Remember to deburr the holes when finished. This piece now forms the upper blade of the brake.

Place a valve spring on each of the spring perches. The springs should be placed so the 3/8" hole is not covered. Place the upper blade in the brake to rest on the springs. Orient the upper blade so the face with the holes is the leg which is resting on the springs. Place a 2 1/2 inch long 3/8" shouldered bolt through the upper blade. Thread a 3/8" nut on the bolt approximately 1/2 inch. Place the end of the bolt through the hole in the spring perch. Screw the 3/8" bolt down until the head just touches the upper blade. Place a 3/8" nut on the end of the bolt sticking through the spring perch. Tighten the nut against the bottom of the spring perch.

Step thirteen

Step 14 T handles

Lay out and cut two 6" lengths of 3/8" round stock. With a grinder, round both ends of the *4" lengths.*

With a grinder, make a small notch in the top of a three inch long 3/8" bolt. The notch should be just deep enough to keep the *6" length* from rolling off the top of the bolt. Repeat the process for another three inch long 3/8" bolt.

Prepare to weld one of the *6" lengths* to one of the 3/8" bolts by placing the round stock in the notch on top of the bolt. Orient the bolt to be in the middle of the round stock. The placement should form a "T" shape.

Weld the round stock to the 3/8" bolt with 1/8" 6011 welding rod at 65 amps. Repeat the placement and welding process for the remaining 3/8" bolt and round stock. These pieces now form the "T" handles for the top blade.

Step thirteen

Step 15

Thread two 3/8" nuts on one of the "T" handles. Position the nuts to be slightly less than 2" apart. With a "C" clamp, place the "T" handle at the connection of the *60 1/8" length* of 2" channel and the end plate. Orient the 3/8" nuts to contact both the 2" channel and the end plate.

Weld the 3/8" nuts to the end plate and 2" channel with 1/8" 6011 rod at 65 amps. Repeat the process for the remaining "T" handle at the other end of the brake assembly. Remove any slag and inspect the welds.

If the "T" handle does not turn freely, remove it from the brake assembly. Thread a 3/8" tap through the upper nut and down through the lower nut. Replace the "T" handle and check to see that it turns freely.

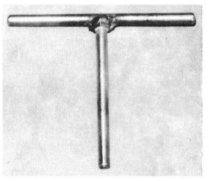

Step fourteen

Step 16 Brake Handle

Finally, we need a handle for the 60" length of 3 1/2" angle iron. To hand fabricate a handle, lay out and cut a 20" length of 1" tubing. Grind one end smooth or weld a 1 1/4" ball bearing on the end.

Flatten the connecting end of the handle to fit against the inside face of 3 1/2" angle iron. With a "C" clamp, attach the handle to the lower face at one end of the angle iron. You may find it convenient to place a second handle on the opposite end of the 3 1/2" angle iron.

Weld the handle to the angle iron with 1/8" 6013 welding rod at 120 amps.

Step 17

A coat of paint will provide a nice finishing touch.

Step fifteen

Step sixteen

Operating Instructions

It is best to experiment with this machine before placing it in service. Begin by laying out a line on a piece of scrap sheet metal. Place the sheet metal under the upper blade and align the layout line with the edge of the upper blade. Lock the blade on the work with the "T" handles. Pull the handle upward to rotate the 3 1/2" angle. This will bend the work on a 3/16" radius beginning at the layout line.

Operating Tips

This brake will bend 5 foot widths of 14 gauge mild steel, 16 gauge stainless steel, and .120" thick aluminium. It will, of course, bend shorter widths of the same materials. Narrow widths of thicker materials can also be bent with this machine, for example 2" x 1/8" cold rolled flat stock. To bend thicker material with this machine, place the work at one end of the blade directly below the "T" handle. Bending thick materials in the middle of the brake will warp the top blade.

When metal is bent it has a tendency to work harden. Prehardened materials have a tendency to crack or break when bent. Therefore, softer metals bend more successfully than harder metals. The hardness of metals can be determined by the alloy or a relative hardness number. A good steel alloy for bending is "1020 cr." This is a mild steel that bends easily without cracking. Two aluminium alloys that work well are "3003" and "6061." The alloy 3003 H-14 is very good for fabrication. The "H-14" indicates that the metal is half hardened and will reach ultimate hardness from working. The 6061 family is tempered for hardness. A T-6 after the alloy number indicates that it is fully hardened. With this alloy, it is best to stay in the T-0 to T-4 range of hardness. Each material will exhibit different bending characteristics so it is best to experiment first

CHAPTER SIX

FLAME CUTTER

Materials List

9 ft. 1 1/4" X 1 1/4" 16 ga. square tubing
5 ft. 2" X 2" X .120 square tubing
3 in. 1" x 1/8" flat stock
36 in. 3/8" all-thread
12 ea. 3/8" nuts
4" X 4" X 1/4" plate
2 ea. 5/16" U bolts
4 ea. 5/16" nuts and flat washers

4 ea. 200SS BARDEN or 200PP FAFNER
    or equivalent ball bearings
12 volt motor
12 volt power supply

Assembly Instructions

Step 1 Mast

Lay out and cut the following lengths of 2" square tubing:

1 @ 36 inches

1 @ 24 inches

Remember to make allowances for the width of the saw blade.

Step 2

Lay out and cut a 45 degree angle on one end of the *24" length* and *36" length* of square tubing. These angles may be laid out with a protractor-head square or by measurement. To lay out by measurement, on one edge scale down 2" from the end and draw a line to the end corner of the opposite edge.

Step 3

Lay out a line 1" from the square end of the *24" length.* This line should be on the longest face of the tubing. Now lay out a centerline on the tubing to cross the previous line. With a center punch and hammer, make a dimple at the intersection of the two lines. If you do not have access to a drill press repeat the layout on the opposite face of the square tubing.

Drill a 3/8" hole at the dimple.

With a drill press, start by anchoring the work to the table so that when the drill penetrates the bottom side of the work it doesn't drill a hole in the table or vise. With a center drill, lightly press the bit on the work and quickly remove it. Check the resulting impression to see if the hole has started in the proper place. With the drill started in the correct position, drill through the work. When finished, deburr the edges of the holes with a deburring knife or round file.

Without a drill press, drill a pilot hole at each dimple on both sides of the square tubing. Use the dimples to guide the drill bit. Change to the larger bit and drill the pilots to the larger size. Deburr the holes with a knife or round file when finished.

Step 4

Prepare to weld the *24" length* and *36" length* of square tubing by placing the ends with the 45 degree angle together. With these pieces properly positioned they should form an "L" shape. With a framing square, check to be sure your placement is square.

Tack weld one corner of this connection. A tack weld is a small weld used to hold position before welding solid. Being careful not to break the tack weld, turn the work over. Check for squareness again, and then tack weld the opposite corner. Should the connection "pull" out of square from the tack weld, lightly tap the pieces back to square with a hammer.

Step four

Weld all four sides of this connection to make it solid. Use 1/8" 6013 welding rod and set the amperage at 105. The amperage may need to be altered if you're not getting good penetration. The work can be placed on its sides or back to facilitate flat horizontal welds, which are easier than vertical or overhead welding. Allow the weld to cool for a few minutes, then chip the slag away and inspect the weld. Any gaps or holes in the weld should be rewelded. These pieces will form the mast when completed.

Step 5

Lay out and cut the following lengths of 1 1/4" square tubing:
> 2 @ 19 inches
> 2 @ 16 3/4 inches
> 1 @ 10 1/2 inches
> 1 @ 8 inches
> 2 @ 7 inches
> 2 @ 2 inches

Start with the longest length, then the next longest and so on. By doing this, if you make a mistake the piece can be used somewhere other than the scrap pile.

Step 6

Lay out a centerline on both *2" lengths* of 1 1/4" square tubing. Then on each piece lay out a line 1" from the end to cross the centerline. Mark the ends the measurements were taken from for later reference. With a center punch and hammer, very lightly make a dimple at the intersection of the lines. This tubing is thin and will bend if struck hard with the hammer and center punch. Be careful not to bend the tubing.

Drill a 3/8" hole though the *2" lengths* of 1 1/4" square tubing at each of the dimples. Remember to deburr the holes when finished.

Step 7

Lay out and cut the following lengths of 3/8" all-thread:
> 1 @ 13 1/2 inches
> 1 @ 11 1/2 inches
> 1 @ 11 inches

Remember to cut the longest length first and so on. When completed, carefully sand or grind the ends to remove any burrs that would prevent a nut from being threaded on the pieces.

Step 8

Lay out four lines on the 36" length of the mast. These lines should be on the short face of the 36" length. Measure from the bottom face of the 24" length to lay out these lines. The lines should be laid out at 7", 8 1/4", 18 1/4", and 19 1/2". Now lay out a centerline to cross each of these lines.

Prepare to weld the *2" lengths* of 1 1/4" tubing to the mast by orienting the 36" leg of the mast flat and the 24" leg pointing upward. Place one of the *2" lengths* on the mast between the 7" and 8 1/4" layout lines. Place the remaining *2" length* on the mast between the 18 1/4" and 19 1/2" layout lines. Orient the *2" lengths* so the end with the reference mark is resting on the mast. Also orient the holes toward each other and place the 13 1/2" length of all-thread through the holes. Align the centerlines on the 2" lengths with the centerline on

Step eight

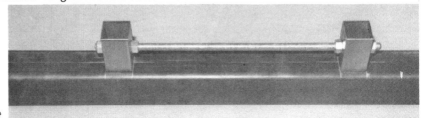

Step eight

the mast. Clamp both 2″ lengths to the mast with "C" clamps. Tack weld opposite corners of both 2″ lengths and double check the alignment, then remove the all-thread and clamps.

Welding the 1 1/4″ square tubing with an AC welder may be difficult. The 16 ga. wall thickness has a tendency to melt and fall away when trying to weld on it. It would be best to use a TIG or MIG welder for the 1 1/4″ tubing, but an AC welder with AC welding rod will work. Weld the *2″ lengths* to the mast with 3/32″ 7014 welding rod with the amperage set at 45. Remove any slag and inspect the welds.

Step 9

Lay out and cut a 4″ x 4″ piece of 1/4″ plate if you don't already have one. With a grinder, straighten and smooth the edges of the plate.

Lay out a line 1/2″ from the edge, and parallel with the edge, on all four sides of the 1/4″ plate. With a center punch and hammer, make a dimple at the four intersections of the layout lines. When completed this piece will be the anchor plate.

Drill a 3/8″ hole at each of the dimples on the anchor plate.

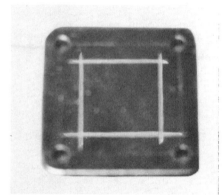

Step ten

Step 10

Lay out a line 1″ from the edge, and parallel with the edge, on all four sides of the anchor plate.

Prepare to weld the anchor plate to the mast by positioning the end of the 36″ leg in the center of the layout lines on the anchor plate. Tack weld opposite corners of the connection. With a combination square, check to be sure the connection is square.

Weld the anchor plate solidly to the mast with 1/8″ 6013 welding rod at 105 amps. Remove any slag and inspect the weld. The mast can now be bolted to a welding table or cutting grate.

Step ten

Step 11 Inner Arm

Lay out a line 1 5/8″ from each end of both *7″ lengths* of 1 1/4″ square tubing. Lay out a centerline on both ends of the *16 3/4″ lengths*. Check to be sure the 16 3/4″ lengths are identical in length; if not, grind the excess lengths from the longer piece.

Place one of the *7″ lengths* and *16 3/4″ lengths* together aligning the centerline of the 16 3/4″ length with either layout line on the 7″ length. With a combination square, check the squareness of the placement. Tack weld opposite corners of this connection to hold its position. Align the remaining *16 3/4″ length* with the other layout line on the 7″ length. Check the squareness and tack weld into position. Place the remaining *7″ length* across the open ends of the *16 3/4″ lengths*. Align the centerlines of both 16 3/4″ lengths with the layout lines on the 7″ length. Measure the distance between the ends of the 7″ lengths. The distance between the end should be the same. This is more important than the squareness of the 16 3/4″ and 7″ pieces. With the pieces properly aligned, tack weld opposite corners of both *16 3/4″ lengths* to the *7″ length*.

Weld all sides of these four connections with 3/32″ 7014 rod at 45 amps. Remove any slag and inspect the welds. This assembly will be used as the inner arm of the panograph section of the flame cutter.

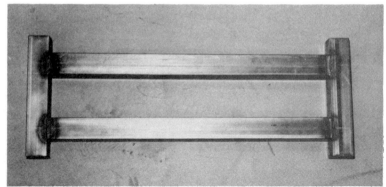

Step eleven

Step 12

Prepare to insert the bearings into the inner arm by standing the arm upright with the open ends of the 7" lengths upward. The bearings have an outside diameter larger than the opening in the end of the square tubing, therefore, the bearing must be driven into the tubing. Start by placing the ball end of a ball peen hammer in the open end of one 7" length. The ball should be larger than the opening in the tubing to work properly. With another hammer, lightly strike the face of the ball peen hammer. This should cause a slight flair in the end of the tubing. The flair needs to be just large enough to start the bearing into the tubing. Use a socket, with a diameter slightly smaller than the diameter of the bearing, and a hammer to drive the bearing into the tubing. Be careful to use a socket that will drive the bearing by pressing the outer bearing race and not the seal. Drive the bearing into the tubing until the bearing is flush with the end of the tubing. Repeat this process for each of the remaining bearings and open ends of tubing.

Step twelve

Connect the inner arm to the mast with the 13 1/2" length of all-thread. To accomplish this, pass the all-thread through the holes in the bottom 2" length of 1 1/4" square tubing, through the bearings in one end of the inner arm, and through the holes in the upper 2" length of 1 1/4" square tubing. Before passing the all-thread through the upper connection on the mast, place a 3/8" nut on the all-thread and thread it down to the top bearing of the arm. Now lift the arm so the all-thread passes through the upper connection and the lower end of the all-thread passes through the lower connection. Holding this position, place a 3/8" nut on the bottom of the all-thread and thread it up to the bottom bearing of the arm. Also place a 3/8" nut on the top end of the all-thread and thread down until the end of the all-thread is flush with the top of the nut. Now lower the arm so the lower end of the all-thread passes back through the lower connection on the mast. Place a 3/8" nut on the bottom of the all-thread. Adjust the uppermost nut and lowest nut to center the all-thread between the upper and lower connections of the mast. With the all-thread properly positioned, lightly snug the bottom nut with a wrench.

Step 13 Outer Arm

Lay out and cut opposing 45 degree angles on the ends of the 10 *1/2" length* of 1 1/4" square tubing. These angles may be laid out with a protractor-head square or by measurement. To lay out by measurement, on one edge scale down 1 1/4" from the end and draw a line to the end corner of the opposite edge. On the same side (edge) that you measured the 1 1/4", repeat the layout for the other end.

Step twelve

Step 14

Lay out and cut a 45 degree angle on one end of both *19" lengths* of 1 1/4" square tubing. These angles may be laid out with a protractor-head square or by the measurement method.

Lay out a line 1" from the end opposite the 45 degree angles for both pieces. This line should be on the longest face of the tubing. Now lay out a centerline to cross the previous layout line. With a center punch and hammer, very lightly make a dimple at the intersection of the lines. Remember this tubing is thin and will bend if struck hard with the hammer and center punch.

Drill a 3/8" hole at each of the dimples on the *19" lengths* of 1 1/4" square tubing.

Step 15

Lay out a line 7" from the square end of one *19" length* of 1 1/4" square tubing. The layout line should be on a face with a 45 degree angle on the end. Lay out the remaining *19" length* of square tubing in the same manner as before but the layout should be on the opposite face with a 45 degree angle.

Lay out a centerline on both ends of the *8" length* of 1 1/4" square tubing.

Prepare to weld one of the *19" lengths* and the *8" length* by placing them together, aligning the layout line on the 19" length with the centerline of the 8" length. The *8" length* should be contacting the short face of the *19" length*. With a combination square, check the squareness of the placement. Tack weld opposite corners of this connection.

Step fifteen

Prepare to weld the remaining *19" length* to the 8" length by positioning the pieces together as before and place the *11 1/2" length* of all-thread through the 3/8" holes of both 19" lengths. Check the squareness and tack weld opposite corners of the connection.

Prepare to weld the *10 1/2" length* to the *19" lengths* by placing the pieces together matching the 45 degree angles at the ends of each piece. With a framing square, check the squareness of the placement. Tack weld opposite corners of both connections and remove the *11 1/2" length* of all-thread from the assembly.

Weld all sides of the four connections of this assembly with 3/32" 7014 rod at 45 amps. Remove any slag and inspect the welds. This assembly will be used as the outer arm of the panograph section of the flame cutter.

Step 16 Motor Mount

This piece will need special attention. The mounting bracket design will depend on your specific motor. This machine requires a motor that has a gear reduction. Ford wiper motors or power window motors are a good source of such motors. A quick search of the local wrecking yard should turn up a motor suitable for this use. The important thing is to have a motor that will run at a speed slow enough for the cutting torch to cut the material.

Step sixteen

The motor mount must fasten the motor to the outer end of the outer arm. The motor should be located such that the drive shaft extending from the motor is approximately 1 inch from the end of the arm and 1 inch above the arm.

Weld the motor bracket to the outer arm with 3/32" 7014 rod at 45 amps. Remove any slag and inspect the weld.

Step 17

Prepare to weld one of the "U" bolt brackets to the arm by standing the arm upright. The *19" length* with the motor bracket should be oriented upward. Place the "U" bolt bracket on the upper face of the bottom 19" length, approximately 1/2" from the 10 1/2" length. Tack weld the bracket into position.

Weld the bracket to the arm with 3/32" 7014 welding rod at 45 amps.

NOTE

This flame cutter uses your cutting torch. All cutting torches are not the same, therefore, careful planning is needed at this point. A centerline through the tip of your torch needs to pass through the shaft extending from the drive motor. This alignment is necessary for the torch to trace an exact outline of the cutting template. The torch handle is attached to the outer arm with the "U" bolts. The location of the second "U" bolt will depend on the length of the torch handle.

Place the tip end of the torch in a "U" bolt. Pass the ends of the "U" bolt through the bracket on the bottom of the arm. Check the alignment of the tip with the 1/4" hole in the motor bracket. Now lay out a location for the second "U" bolt. The bolt should be

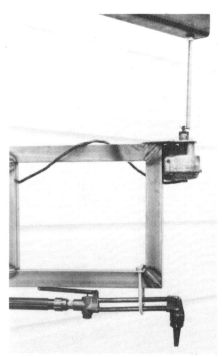

The centerline of your torch tip must pass through the center of the shaft extending from the motor drive.

Layout of torch in "U" brackets. Be sure bracket will not interfere with torch valves.

positioned near the hose end of the torch. Check to be sure the bolt will not interfere with the valves. When the layout is complete remove the torch from the arm and tack weld the second bracket in position.

Weld the second "U" bolt bracket to the arm with 3/32" 7014 welding rod at 45 amps.

Step 18

Attach the torch to the arm as before. On the lower *19" length* mark the location where your thumb would normally operate the cutting lever of the torch. Remove the torch and lay out a centerline to intersect with the previous marking. With a center punch and hammer, make a dimple at the intersection of the lines.

If you have a tap and die set, drill a 1/4" hole using the dimple as a guide for the drill bit. Drill through the tubing. Once the hole is drilled, use a 5/16" tap to thread the hole.

If you do not have a tap and die set, drill a 5/16" hole instead. Place a 5/16" bolt through the hole from the bottom side of the tubing and put a 5/16" nut, finger tight, on the bolt. Weld two sides of the nut to the top surface of the tubing and remove the bolt.

Place a 5/16" bolt through the lower leg of the outer arm. Thread the bolt through the tubing until the bolt just passes through the tubing.

Step 19

Connect the outer arm to the inner arm with the 11 1/2" length of all-thread. To accomplish this, pass the all-thread through the holes in the bottom length of square tubing in the outer arm, through the bearings in the end of the inner arm, and through the holes in the upper length of outer arm. Before passing the all-thread through the upper connection of the outer arm, place a 3/8" nut on the all-thread and thread it down to the top bearing of the inner arm. Now hold the nut and twist the all-thread so it passes through the upper connection approximately 1 3/4". Holding this position, place a 3/8" nut on the bottom of the all-thread and thread it up to the bottom bearing of the inner arm. Also place a 3/8" nut on the top of the all-thread and thread the nut down until the end of the all-thread is flush with the top of the nut. Now feed the all-thread down so the all-thread passes back through the lower connection on the outer arm. Place a 3/8" nut on the bottom of the all-thread. Adjust the uppermost nut and lowest nut to center the all-thread between the upper and lower connections of the outer arm. With the all-thread properly positioned, lightly snug the bottom nut with a wrench. Use the 3/8" nuts adjacent to the bearings to center the inner arm between the connections of the outer arm. With the inner arm properly positioned, lightly snug both nuts with a wrench.

Connect the motor to the motor bracket on the outer arm. Connect a wire to the motor and run the wiring along the top of both arms and down the mast to the base plate. Connect the wires at the base plate to the 12 volt power supply.

Attach the torch to the outer arm with the "U" bolts. Check the for proper alignment of the cutting tip with the drive motor.

Place a 3/8" nut on on end of the *11" length* of all-thread and thread the nut down

Step nineteen

approximately 2 1/2" form the end. Insert the 2 1/2" portion of the all-thread through the 24" leg of the mast from the bottom. Place a 3/8" nut on the end of the all-thread extending through the top of the mast. With a wrench, tighten the top nut on the mast. Place two 3/8" nuts on the lower end of the all-thread.

Step 20

A coat of paint provides a nice finishing touch to the machine.

Operating Instructions

This machine traces a pattern and uses your torch to cut a replica of the pattern. The patterns can be formed from cardboard or metal. Patterns are attached to the all-thread extending down from the top of the mast. The drive motor traces the pattern while the torch is cutting to form the replica.

The power supply must be a 12 volt D.C. type. A variable resistor must also be used in order to control the motor speed. Toy train transformers are a good power supply because they have a variable resistor built into the unit. The speed of the torch feed is controlled from the 12 volt power supply. When the power is turned up, the drive motor runs faster and increases the feed speed.

The 5/16" bolt in the outer arm is used to control the cutting lever as your thumb would if holding the torch in hand.

The thickness of material that can be cut with this machine will depend on your torch. Different material thickness will, however, require adjustment of the machine. The inner arm has approximately 2" of vertical adjustment where it connects to the mast. The distance between the cutting tip and the work can be set with the nuts on the 13 1/2" length of all thread.

Operating Tips

This machine will cut a maximum straight line of 60 inches. It will also cut a circle with a maximum diameter of 34 inches. Any shape that is smaller than the maximum limits may also be cut.

Once a height adjustment of the inner arm has been established for a particular thickness of material, it will be helpful to mark the adjustment location on the mast for later reference.

Notes On Operation

The templates should be drilled and placed on the 3/8" All-Thread extending down from the mast. It should be "stationary" between two 3/8" nuts (one on top and one on the bottom). The motor shaft needs a magnetic tracer to track the template (straight or irregular shaped). Williams Low Buck Tools is one source (see note at upper left of this page). Another manufacturer is ESAB. Most welding supply shops that carry Airco products carry ESAB products.

Regarding templates, obviously, if you use a cardboard, wood or aluminum template you will have to hold the tracer against the template. The advantage of a cardboard template is the ease and speed of creating the template. Cardboard templates can be used when you need only one cut out of an unusual or particularly hard shape. Typically, when fabricating you will need 4 or 8 of something. The thing to do here is fabricate the first one, use it as a template and then use the flame cutter to make the others. Sheet metal templates are not recommended because the tracers don't always stick and it is nearly as much work to make a sheet metal template as the actual part.

CHAPTER SEVEN
CHASSIS STAND

Materials List

26 ft. - 2″ X 2″ X .120″ square tubing
8 in. - 3 1/2″ I.D. X 3/16″ pipe
16 in. - 3″ I.D. X 3/16″ pipe
12″ X 8″ X 1/4″ flat plate
2 ea. - 3/8″ X 1″ bolt

8 ea. - 1/2″ X 3″ U bolt
8 ea. - 1/2″ nut
8 ea. - 1/2″ flat washers

Assembly Instructions

Step 1 Base/Stand

Lay out and cut the following lengths of 2" square tubing:

1 @ 54 inches
1 @ 42 inches
2 @ 30 inches

Start with the longest length, then the next longest and so on. By doing this, if you make a mistake the piece can be used somewhere other than the scrap pile. Remember to make allowances for the saw blade width.

Step 2

One end of the *42" length* needs to be "fish-mouthed" for the 3 1/2" pipe. A fish-mouth is a notch where something round, like pipe, fits on another member. A fish-mouth can be laid out by using one end of the pipe as a template. To do this, place the pipe so the arc described by the pipe runs from one edge to the other edge and meets the corners at the end. With the grinder, remove the circular segment on both sides of the tubing to form the fish-mouth. On the other end of the 42" length lay out a centerline. This line should be on the same face as one of the fish-mouths.

Step 3

On each of the *30" lengths* of 2" square tubing, lay out a line at 15" from the end.

On the *54" length* of 2" square tubing, lay out a centerline (approximately one inch long) at both ends. A centerline marks the center axis of a member. In this case, it would be drawn one inch from either side and run long ways on the tubing.

Prepare to weld one of the *30" lengths* and the *54" length* together by aligning the centerline of the *54" length* and the line at 15" on the *30" length*. With a framing square, check to be sure your placement is square. Tack weld one corner of this connection. A tack weld is a small weld used to hold position before welding solid. Being careful not to break the tack weld, turn the work over. **Check for squareness** again, and then tack weld the opposite corner. Should the connection "pull" out of square from the tack weld, lightly tap the pieces back to square with a hammer.

Prepare to weld the remaining *30" length* and the *54" length* together by aligning the centerline on the opposite end of the *54" length* and the line at 15" on the *30" length*. With a framing square, check to be sure your placement is square. With the pieces properly positioned, tack weld this connection.

Weld all four sides of both connections to make them solid. Use 1/8" 6013 welding rod and set the amperage at 105. The amperage may need to be altered if you're not getting good penetration. The work can be placed on its sides or back to facilitate flat horizontal welds, which are easier than vertical or overhead welding. Allow the weld to cool for a few minutes, then chip the slag away and inspect the weld. Any gaps or holes in the weld should be rewelded. With these pieces connected, this now forms the base.

Step 4

Lay out a line 29" from either end of the 54" length. This line should be drawn on one of the side surfaces and not the top.

Prepare to weld the *42" length* to the base by placing it on the 54" length and aligning the centerline of the 42" length and the line at 29" on the 54" length. The fish-mouth should also be aligned with the 30" lengths. With a framing square, check for squareness between the 54" and 42" lengths.

Tack weld one corner of the connection, check the squareness and tack weld the opposite corner. Weld this connection solid and inspect as on the previous connection. This connection forms the upright.

Step 5

Lay out and cut a *4" length* of 3 1/2" pipe if you don't already have it. On this piece, lay out a short line 2" from either end. With a center punch and hammer make a dimple in the line.

If you have a tap and die set, drill a 5/16" hole using the dimple as a guide for the drill bit. Once the hole is drilled use a 3/8" tap to thread the hole. If you thread the hole with a tap, go to step 6; if not, continue to the next paragraph.

If you do not have a tap and die set, drill a 3/8" hole instead. Place a 3/8" nut on a 3/8" bolt. Place the bolt through the hole from the outside of the pipe. Put a 3/8" nut, finger tight, on the bolt where it extends inside the pipe. Weld two sides of the nut which is on the outside of the pipe. Remove the nut inside the pipe and the bolt.

Step 6

Place the *4" length* of 3 1/2" pipe on the fish-mouthed upright. Orient it so that both ends extend 1" beyond the upright, and the bolt hole is upward. The centerline of the pipe should be parallel with the 30" legs of the base.

Tack weld opposite corners of the fish-mouth to hold the position. With the pipe tacked in place, weld all four sides of this connection to make it solid. Use 1/8" 6013 welding rod and set the amperage at 105. The amperage may need to be altered if you're not getting good penetration.

Step 7 Head Plate

Lay out and cut a 6" x 8" piece of 1/4" plate if you don't already have it. Use a grinder to round off any sharp corners. Lay out a line parallel with the edge so that the line is 1 1/2" from the edge. Repeat the layout for the three other sides. With a center punch, make a dimple at each intersection of the four lines.

Drill four 1/2" holes using the dimples to guide the drill. With a file or deburring knife, clean the edges of the holes.

Lay out the center of the *6" x 8" plate* by drawing a line 4" from one edge in the long direction, then drawing a line 3" from an adjacent edge. Lay out a mark 1 1/2" from the center on each line. This will help locate the 3" pipe in the middle of the plate.

Step 8

Lay out and cut a *8" length* of 3" pipe if you don't already have it.

Place the *8" length* in the middle of the plate by using the layout marks. Tack weld opposite sides of the pipe to prevent moving, then weld solid using 1/8" 6013 welding rod at 120 amps. This now forms the head.

Place the 3" pipe of the head plate, inside the 3 1/2" pipe of the stand and lock in position with a 3/8" bolt.

Step 9

Place two U bolts through the head plate for later use. Place a flat washer and 1/2" nut on both ends of the U bolts.

Step 10

Two chassis stands are necessary for a single car. Step 9 completed one stand. Return to step 1 and repeat the instructions for the second stand.

With both stands constructed you can now paint them for a completed project.

Operating Instructions

These stands are primarily used for chassis construction. After the frame, roll cage, etcetera, have been tack welded together, the assembly is attached to the chassis stands. With a chassis attached to these stands it may be rotated to facilitate flat or horizontal welds, which are easier than vertical or overhead welding.

Place one stand at each end of the chassis. Lift one end of the chassis with an engine hoist to allow the U bolts to attach the chassis to the stand. Finger tighten the nuts on the U bolts, allowing the chassis to be lifted at the other end. Hoist the opposite end of the chassis and attach it to the other stand. Before releasing the chassis from the hoist **be sure** the locking bolts in each upright are tight. With the chassis attached to the stands, it may now be rotated for easy welding.